Service Provision and Rural Sustainability

Access to quality services and community infrastructure are vital parts of supporting sustainable and resilient rural and small town places. Renewing outdated infrastructure and supporting the delivery of services in rural communities present significant challenges from the constrained fiscal and policy realities of the 21st century.

Drawing upon contributors from five Organisation for Economic Co-operation and Development (OECD) countries, this book describes innovative service delivery and community infrastructure models that are appropriate to the contemporary rural and resource-dependent regions of developed economies. The examples show that an entrepreneurial approach to service delivery and infrastructure provision by local organizations and governments is needed. Critical economic and community development supports are crucial to assist creative and innovative sets of solutions that work for small communities. Chapters in this book argue that community development foundations for resilient rural and small town communities and regions must be co-constructed and co-delivered in partnership by both local and senior government actors, in terms of both policy and committed resources.

This volume will be extremely valuable for students, scholars, and community development practitioners exploring policy-making, government initiatives, and community service provision in rural and small town places.

Greg Halseth is the Canada Research Chair in Rural and Small Town Studies in the Geography Program at University of Northern British Columbia, Canada.

Sean Markey is a professor with the School of Resource and Environmental Management at Simon Fraser University, Canada.

Laura Ryser is the research manager of Rural and Small Town Studies in the Geography Program at University of Northern British Columbia, Canada.

Perspectives on Rural Policy and Planning

Series Editors: Andrew Gilg and Mark Lapping

This well-established series offers a forum for the discussion and debate of the often-conflicting needs of rural communities and how best they might be served. Offering a range of high-quality research monographs and edited volumes, the titles in the series explore topics directly related to planning strategy and the implementation of policy in the countryside. Global in scope, contributions include theoretical treatments as well as empirical studies from around the world and tackle issues such as rural development, agriculture, governance, age and gender.

For more information about this series, please visit: www.routledge.com/Perspectives-on-Rural-Policy-and-Planning/book-series/ASHSER-10354

Service Provision and Rural Sustainability

Infrastructure and Innovation

**Edited by Greg Halseth, Sean Markey and
Laura Ryser**

Routledge
Taylor & Francis Group

LONDON AND NEW YORK

First published 2019
by Routledge
2 Park Square, Milton Park, Abingdon, Oxon OX14 4RN

and by Routledge
52 Vanderbilt Avenue, New York, NY 10017, USA

First issued in paperback 2020

Routledge is an imprint of the Taylor & Francis Group, an informa business

British Library Cataloguing-in-Publication Data
A catalogue record for this book is available from the British Library

Library of Congress Cataloging-in-Publication Data
Names: Halseth, Greg, editor. | Markey, Sean Patrick, 1970- editor. | Ryser, Laura, editor.
Title: Service provision and rural sustainability : infrastructure and innovation / [edited by] Greg Halseth, Sean Markey and Laura Ryser.
Description: Abingdon, Oxon ; New York, NY : Routledge, 2019. | Includes bibliographical references and index.
Identifiers: LCCN 2018031087| ISBN 9781138483729 (hbk : alk. paper) | ISBN 9781138483736 (ebk : alk. paper) | ISBN 9781351054010 (mobi/kindle)
Subjects: LCSH: Rural development. | Community development, Rural. | Sustainable development. | Municipal services. | Infrastructure (Economics)
Classification: LCC HN49.C6 S4635 2019 | DDC 307.1/412--dc23
LC record available at https://lccn.loc.gov/2018031087

ISBN 13: 978-0-367-58358-3 (pbk)
ISBN 13: 978-1-138-48372-9 (hbk)

Typeset in Times New Roman
by Integra Software Services Pvt. Ltd.

Our friend Don Manson,
Your continuing commitment to rural and small town communities
inspires us and our work

Contents

Figures

Tables

Contributors

Joshua Barrett, Research Associate, Memorial University Joshua Barrett is a senior policy analyst in the Department of Tourism, Culture, Industry, and Innovation within the Government of Newfoundland and Labrador. In this current role, he works to shape public policy related to innovation and regional development. His research interests include economic geography, regional development, public policy, democracy, governance, and labour mobility. Joshua holds a Bachelor of Arts in geography and political science, and a Master of Arts in geography from Memorial University of Newfoundland.

Greg Blackburn, General Manager, IMC (UK) Learning Ltd. Greg Blackburn is the general manager for IMC (UK) Learning Ltd. With a PhD from City, University of London, his research interests include change management, information systems, and educational technology, leading to research activity and publications in two distinct fields: 'public administration' and 'e-learning'. Greg conducted the first post-implementation review of Australia's first whole-of-government initiative, Service Tasmania. Recently, his research focuses on understanding how educational psychology and technology can benefit problem-solving and critical thinking skills development. He specializes in transformative learning and development in delivering innovative and high-value corporate learning.

Sarah-Patricia Breen, Post-doctoral Fellow, Conservation of Change Lab, University of Saskatchewan Sarah-Patricia Breen is a post-doctoral scholar with the Conservation of Change Lab at the University of Saskatchewan. Sarah holds a PhD in resource and environmental management from Simon Fraser University, a master's in geography from Memorial University, and a BA (honours) in geography from Lakehead University. Her work experience includes the public and private sectors, as well as academia. Her research interests include water management, climate change adaptation, regional resilience, and all things rural. Sarah currently serves as president of the Canadian Rural Revitalization Foundation board of directors.

Sean Connelly, Senior Lecturer, Department of Geography, University of Otago, New Zealand Sean Connelly is a senior lecturer in the Department of

Geography at the University of Otago. His research focuses on sustainable communities, particularly on how communities mobilize resources and build capacity to disrupt the status quo. Sean examines these processes in the context of regional development and the transition to sustainability, alternative food networks and food system sustainability, and the convergence of social economy and sustainable community development movements. He is a founding member of Our Food Network Dunedin.

Brian Dollery, Professor, UNE Business School, University of New England Brian Dollery is a professor of economics and the director of the Centre for Local Government at the University of New England, as well as visiting professor at the Faculty of Economics at Yokohama National University. He has worked with local government across Australia and New Zealand, largely in the area of structural change and financial sustainability. Recent books include *Funding the Future* (2013), *Councils in Cooperation* (2012), *Local Government Reform* (2008), *The Theory and Practice of Local Government Reform* (2009), *Reform and Leadership in the Public Sector* (2007), and *Australian Local Government Economics* (2006).

Ryan Gibson, Libro Professor of Regional Economic Development, School of Environmental Design and Rural Planning, University of Guelph As the Libro Professor of Regional Economic Development in the School of Environmental Design and Rural Planning at the University of Guelph, Ryan's research focuses on rural development, governance, philanthropy and wealth, and public policy. Recent publications include *Place Peripheral: Place-Based Development in Rural, Island, and Remote Communities* (2015) and *Building Community Resilience* (2017). He is the president of the Canadian Community Economic Development Network, past president of the Canadian Rural Revitalization Foundation, and chair of the Institute of Island Studies Advisory Board.

Greg Halseth, Professor and Canada Research Chair in Rural and Small Town Studies, Geography Program, University of Northern British Columbia, Canada Greg's research examines rural and small town community development, and community strategies for coping with social and economic change in northern BC. He is also the co-director of UNBC's Community Development Institute. Greg has served on the governing council of the Social Sciences and Humanities Research Council of Canada, the Advisory Committee on Rural Issues for the Federal Secretary of State for Rural Development, and the Community Advisory Committee for the BC Ministry of Forests Mountain Pine Beetle Task Force.

Neil Hanlon, Professor, Geography Program, University of Northern British Columbia, Canada Neil Hanlon is a professor of geography at the University of Northern British Columbia, Canada. His research interests include rural health service care delivery, community adaptations to social and economic change, and regional health governance. His work appears in journals such as *Health and Place* and *Social Science and Medicine,* and he recently co-edited with Mark Skinner (Routledge, 2016) a book about the role of the voluntary sector in making resource-dependent communities more supportive of those aging in place.

Michael Hynes, School of Political Science and Sociology, NUI Galway, Ireland Michael Hynes is a lecturer in political science and sociology, specializing in environmental sociology, mobilities and sustainable transport, sustainability research, and society-technology-environment interactions. He is co-chair of the Governance and Sustainable Development research cluster, a member of the Teaching and Learning Committee, and a board member of the Social Sciences Research Centre. His research for the ConsEnSus Project (2009 to 2013) critiqued telework policy and practice for their overemphasis on technology-induced efficiency and productivity gains and identified contradictions with wider social and environmental sustainability goals in the areas of work and consumption.

Wayne Kelly, RPLC Project Coordinator, Brandon University Wayne has been studying rural development for more than 10 years and has received his master's of rural development from Brandon University. He is presently a PhD student at National University of Ireland, Galway. At the Rural Development Institute, Wayne is focused on applying innovative development research to help rural Canada realize the opportunities facing it as society, the economy, and technology change. As part of the RDI team and as the coordinator of the Rural Policy Learning Commons (RPLC), Wayne hopes to explore new and creative areas to support communities, organizations, businesses, and government in rural Canada.

Kaye Knight, PhD, M.Ed, BA App. Sci (Nursing), RN, Manager, Education and Research, Rural Northwest Health Kaye is the manager of education and research for Rural Northwest Health, Victoria, Australia. A public health service funded by the State and Commonwealth Government, Rural Northwest Health is an award-winning rural health service. Kaye's research interests focus on rural health. Kaye has made a conscious decision to remain in the rural health sector to engage clinicians and health providers in research, and to assist with the dissemination of the innovative work being undertaken in the rural health and aged care sectors.

Martha MacLeod, Professor, Schools of Nursing and Health Sciences, University of Northern BC Martha MacLeod is a professor in the schools of Nursing and Health Sciences at the University of Northern British Columbia, where she is the Northern Health – UNBC Knowledge Mobilization Research Chair and co-lead of UNBC's Health Research Institute. Her work examines knowledge creation and translation within health services and health human resources, particularly in rural and northern settings. Martha is engaged in partnered research, and is active in national and regional multidisciplinary research and knowledge translation networks.

Sean Markey, Professor, School of Resource and Environmental Management, Simon Fraser University, Canada Sean is a professor with the School of Resource and Environmental Management and an associate with the Centre for Sustainable Community Development and Department of Geography. His research concerns issues of local and regional economic development, rural and small town development, community sustainability, and sustainable infrastructure. He continues to work with municipalities, non-profit organizations, aboriginal communities, and the business community to promote and develop sustainable forms of community economic development. He serves on the board of directors with the Vancity Community Foundation and the Silva Forest Foundation.

Sarah Minnes, PhD Candidate, Environmental Policy Institute, Memorial University – Grenfell Campus Sarah is a PhD candidate at Memorial University of Newfoundland exploring the role of capacity and governance for source water protection in rural regions in Ontario. As a registered professional planner, she has served as water liaison for Municipalities Newfoundland and Labrador and has been involved in research related to regional planning and development, and sustainable rural drinking water systems at Memorial University. Sarah has also worked for the York Region Forestry, the Ministry of Natural Resources, and the Ontario Federation of Anglers and Hunters with the Bring Back the Salmon program.

Catherine Morley, MBA, Grad. Cert. Gerontology, Grad. Cert. Quality Mgt, RN Catherine is the chief executive officer for the Wimmera Health Care Group, Victoria, Australia. Prior to this, she was the chief executive officer at Rural Northwest Health. Catherine has a strong people focus and a specialization in providing services for people with a disadvantage, and with a commitment to ongoing learning and development and change management. Catherine is an aged-care industry leader, and a hallmark of Catherine's work is supporting innovation in rural health care that is centred around client outcomes and continuous improvement.

Etienne Nel, Professor, Department of Geography, University of Otago, New Zealand Etienne's research focuses on issues of small town development and local, regional, and economic development in southern Africa and in Australasia. He is the managing editor of the *New Zealand Geographer* and commissioning editor for Australasia for *Local Economy*. Recently published books include: *Africa: Diversity and Development* (Routledge); *South Africa: Past, Present, and Future* (Routledge); *Local Economic Development in the Developing World* (Routledge); and *Geographies of Development* (Routledge).

He is the current chair of the International Social Science Council's (UNESCO) Comparative Research Programme on Poverty.

Trish Reay, Professor, Alberta School of Business, University of Alberta, Canada Trish is a professor of strategic management and organization at the University of Alberta School of Business. She also holds a part-time affiliation as professor of entrepreneurship and innovation at the Warwick Business School. She currently serves as editor-in-chief of *Organization Studies*. Her research interests include organizational and institutional change, professions, and professional identity in the context of health care and family firms. Published articles from these research streams appear in *Academy of Management Journal, Organization Studies, Journal of Management Studies, Work and Occupations*, and *Family Business Review*.

Laura Ryser, Research Manager, Rural and Small Town Studies, Geography Program, University of Northern British Columbia Laura Ryser is the research manager of the Rural and Small Town Studies program at the University of Northern British Columbia. She has completed research in communities across northern BC and throughout Canada to explore rural restructuring and community transition, community economic development, labour mobility, industry-community relationships, seniors' needs, the deployment of social capital, and partnership development. Her research interests include small town community change, institutional barriers to change, building resiliency to respond to restructuring trends, smart service and infrastructure arrangements, and rural poverty.

Erin Sherry, Senior Agricultural Economist, Agri-Food and Biosciences Institute, Northern Ireland Erin Sherry (neé Minihan) is a senior agricultural economist at the Agri-Food and Biosciences Institute in Northern Ireland. She conducts a wide range of research for the Northern Ireland Department of Agriculture, Environment and Rural Affairs to support public policy development and serves on the department's Axis 3 working group. Her current work explores economic structure, employment, rural development policy, and rural proofing. Erin also has experience researching environmental issues such as greenhouse gas emissions, carbon sequestration, and land use.

Sally Shortall, Professor, School of Sociology, Social Policy and Social Work, Queen's University Sally is interested in questions on rural women, farming women, rural development theory and practice, governance, community and stakeholder engagement in policy processes, and how evidence is used to inform policy. She is currently exploring the social construction of knowledge to inform rural policy and the power struggles between empirical and normative knowledge.

David Snadden, Professor, Rural Doctors' UBC Chair in Rural Health, University of British Columbia David Snadden is a professor in the

Department of Family Practice at the University of British Columbia, and the Rural Doctors' UBC Chair in Rural Health. He graduated from Dundee University in 1977 and was a rural family physician in the Highlands of Scotland. He came to Canada in 2003 to lead the development of the Northern Medical Program, a distributed site of the Faculty of Medicine at UBC, and from 2011 to 2016 was the executive associate dean of education for the faculty.

Kelly Vodden, Associate Vice-President, Research and Graduate Studies, Memorial University – Grenfell Campus Kelly is the associate vice-president (Grenfell) of research and graduate studies and an associate professor with the Environmental Policy Institute at the Grenfell Campus of Memorial University. She also serves as a research associate and advisor to Municipalities of Newfoundland and Labrador, a board member of the Indian Bay Ecosystem Corporation, and a former board member of the Canadian Rural Revitalization Foundation. Kelly's research and publications relate to governance and sustainable community and regional development, with a focus on rural, coastal, often natural resource-dependent communities.

Wendy Walters, M.Hlth Services – Aged Care, Grad.Dip. Gerontology, RN, Warracknabeal Campus Manager – Aged Care, Rural Northwest Health Wendy is a passionate and innovative manager and advocate for excellent resident care. A public health service funded by the State and Commonwealth Government, Rural Northwest Health is an award-winning rural health service. Wendy is currently the manager of residential aged care services for Rural Northwest Health, Victoria, Australia. She has led innovative, award-winning projects in residential aged care and was the visionary and driver for the Wattle initiative and ABLE-D model of care, both quality improvement and quality of care initiatives in dementia care.

Rachel Winterton, Research Fellow, John Richards Centre for Rural Ageing Research, La Trobe University Rachel's research focuses on how rural communities, governments, and organizations are managing and responding to challenges posed by populations ageing through systems of governance, health, and social infrastructure. She is currently completing a series of projects with international collaborators exploring critical perspectives on volunteering in aging rural communities. Other research interests include the implications of rural retirement migration for rural service provision, rural organizational capacity to facilitate age-friendly communities, and the role of rural systems and structures in facilitating wellness for aging populations.

Acknowledgements

When we began this book, we never could have imagined how fortunate we would be to work with so many colleagues who have a profound and firm commitment to rural communities. We feel fortunate to have collaborated with these scholars and practitioners – a collaboration that has strengthened the insights into processes guiding sustainability and innovation for services and infrastructure in small communities. Our most sincere gratitude, therefore, goes to Erin Sherry, Sally Shortall, Greg Blackburn, Brian Dollery, Etienne Nel, Sean Connelly, Neil Hanlon, Martha MacLeod, Trish Reay, David Snadden, Rachel Winterton, Kaye Knight, Catherine Morley, Wendy Walters, Ryan Gibson, Joshua Barrett, Sarah Minnes, Sarah-Patricia Breen, Kelly Vodden, Wayne Kelly, and Michael Hynes.

Our journey to explore broader changes in rural services began with our involvement in the New Rural Economy (NRE) Project. We feel very fortunate to have been invited to be a part of this project by Bill Reimer of Concordia University. Our colleagues across Canada provided tremendous support for developing a national rural service observatory and beginning to explore issues around innovation. We extend our thanks and appreciation to Bill Reimer (now retired from Concordia University), Tom Beckley (University of New Brunswick), David Bruce (Mount Allison University), Bruno Jean (Université du Québec à Rimouski), Patrice LeBlanc (Université du Québec en Abitibi-Témiscamingue), Doug Ramsey (Brandon University), Ellen Wall (now retired from University of Guelph), the late Derek Wilkinson (Laurentian University), Ivan Emke (Sir Wilfred Grenfell College, Memorial University of Newfoundland), Diane Martz (University of Saskatchewan), Omer Chouinard (Université de Moncton), Anna Woodrow (Concordia University), Steve Plante (Université du Québec à Rimouski), Dianne Looker (Acadia University), Peter Apedaile (now retired from the University of Alberta), and Ray Bollman (Statistics Canada).

Many of the NRE relationships have been extended through the Social Sciences and Humanities Research Council (SSHRC)-funded Rural Policy Learning Commons (RPLC) project, through which we have acquired continued support to expand our networks and research on this important topic. We have been grateful for the continued support of Bill Ashton, Bill Reimer, Ryan Gibson, Philomena de Lima, Sally Shortall, Francesca Regoli, Matteo Vittuari, Wayne Kelly, Michael Blatherwick, and Meghan Wrathall.

Another valued group of Canadian and international colleagues includes fellow researchers interested in rural services, infrastructure, and local governments. They include Marleen Morris (University of Northern British Columbia), Chris Bryant (Université du Montréal), Aleck Ostry (University of Victoria), Mark Skinner (Trent University), Alun Joseph (retired from the University of Guelph), Terri Macdonald (Selkirk College), Sara Boucher (University of Otago), Neil Argent (University of New England), Denise Cloutier (University of Victoria), Malcom Cutchin (Wayne State University), Lars Hallstrom (University of Alberta), Tor-Arne Gjertsen (UiT/The Arctic University of Norway), Rachel Herron (Brandon University), Sarah Lovell (University of Canterbury), Paul Milbourne (Cardiff University), Christine Milligan (Lancaster University), Natalie Waldbrook (Queen's University), Kieran Walsh (NUI Galway), Jeni Warburton (La Trobe University), Elaine Wiersma (Lakehead University), Bojan Fürst (Harris Centre, Memorial University), and Robert Greenwood (Harris Centre, Memorial University).

We have also been fortunate to have wonderful support from many dedicated students over the years, many of whom have pursued careers in rural services and are now strong advocates for sustainable rural communities. These students include Alex Martin, Allison Matte, Chloe Boyle, Amy Gondak, Sarah-Patricia Breen, Karen Heisler, Anisa Zehtab-Martin, Anne Hogan, Ashley Kearns, Brian Stauffer, Carla Seguin, Negar Esfahani, Catherine Fraser, Chelan Zirul, Christine Creyke, Colin McLeod, Courtney LeBourdais, Eric Kopetski, Gretchen Ferguson, Jennifer Crain, Jennifer Herkes, Jenny Lo, Jessica Raynor, Joanne Doddridge, Kelly Geisbrecht, Laura Van de Keere, Lila Bonnardel, Liz O'Connor, Marc Steynen, Melinda Worfolk, Melissa Zacharatos, Michelle White, Mollie Cudmore, Nora King, Onkar Buttar, Pam Tobin, Paul Pan, Priscilla Johnson, Rachael Clasby, Rebecca Goodenough, Rosalynd Curry, Shiloh Durkee, Tobi Araki, and Virginia Pow. Our recent team has included Gerald Pinchbeck, Daniel Bell, Alika Rajput, Alishia Lindsey, Kourtney Cook, Devon Roy, Erin MacQuarrie, Danielle Patterson, Marli Bodhi, Jessica Blewett, and Michael Lait.

A special thanks goes to Lana Sullivan. Lana played a foundational role in establishing the rural and small town studies team at UNBC and worked on its many early projects. Lana expressed a lot of care and heart during our early visits to northern BC's small communities. It created a solid foundation for so much of what we have achieved. Other valued members of our team now include Julia Good (neé Schwamborn), Aita Bezzola, Scott Emmons, and Kyle Kusch, who have provided critical support to bring this collection together.

Additional resources and time for the research and writing that underlies this book come from the federal government's Canada Research Chairs program. Initially in 2002, Greg Halseth was awarded a Canada Research Chair Tier 2 position, and in 2011 he was awarded a Canada Research Chair Tier 1 position. This support is gratefully acknowledged as it allows for Canadian researchers to advance knowledge and understanding in support of their areas of expertise. We also acknowledge funding support from the vice president of research at Simon Fraser University.

Lastly, we pay tribute to our long-term colleague and friend Don Manson. Over the years, he travelled thousands of miles to our small communities, making himself available anywhere and at any time for the people he met. As an original team member of 'the board', he has been an adamant champion for supporting sustainable rural communities, a terrific colleague for so many years, and the host of many 'second breakfasts' along the roads of northern BC.

Perhaps most importantly, we draw our strength from our families. 'Thank you' will never be a large enough expression!

Greg Halseth, Sean Markey, and Laura Ryser
September 2018

Part I

Introduction

1 Introduction

Greg Halseth, Sean Markey, and Laura Ryser

Introduction

Discussion and debate about the future of rural and small town places within developed economies have focused upon the need to create and support more sustainable economies and more resilient communities (Brown and Schafft, 2011; Halseth *et al.*, 2010; Halseth and Ryser, 2018; Markey *et al.*, 2012; OECD, 2010, 2014). In study after study, it is clear that rural and small town places have a promising future in the new global economy, but it is equally clear that poor public and private-sector policy choices, and the application of outdated program and funding solutions, are not supporting this transition to more sustainable economies and resilient communities. As a complement to the literatures on rural and small town transformation, this book devotes its attention to the delivery of needed human services and the infrastructure to support those services.

One of the longstanding and most critical elements to successful rural and small town communities is the availability of an appropriate suite of services and service infrastructure. This edited volume takes up the challenge of human services provision in rural and small town places in developed economies. It calls upon researchers from four OECD states who are experienced in rural and small town services/infrastructure provision and invites them to share critical stories, bound by common themes. The motivation for the volume concerns the continued viability of rural and small town places in a 21st-century political and economic context. The organizing premise is that older models of service delivery, and the supporting infrastructure for that service delivery, are not appropriate in a 21st-century context. As so often happened through the mid-20th century, individual services and the supporting infrastructure of buildings were created as single-purpose entities to deliver only that service in isolation from other services and community needs. Renewing and replacing outdated single-purpose infrastructure and support-ing the human delivery of services in small places are significantly challenged by the fiscal and policy realities of the 21st century.

The chapters in this volume work to identify innovative and creative service delivery models that are appropriate to 21st-century rural and small town places. Each contribution highlights how the case studies advance our understanding of, and the potential for, rural service delivery (processes and/or products). The chapters note how different models and modes of service delivery have emerged out of the challenges confronting more traditional service models. In their case studies, authors share details about the impetus behind new service models or initiatives, as well as the factors that supported or hindered the implementation of the new models. They also discuss the transferability of new models and the roles of various levels of government.

Taken together, the opportunities and challenges of rural service provision within a policy framework marked by 'reactionary incoherence' raise important questions. Throughout this process we have challenged the authors in this volume to address important analytical questions and bring coherence, and a common foundation for discussion, to their contributions. They are also important for readers to keep in mind as they approach different parts of the book. These questions include

- What challenges are facing rural and small town services and service delivery?
- What are the key features that help to identify and define rural and small town services as being 'innovative' and 'creative'?
- What aspects of rural and small town service provision are more suited and appropriate to the realities of the 21st century?
- What if renewed service delivery models are not delivered? What will rural look like then?
- Whose responsibility is it to push for, and then deliver, these new or innovative models of service delivery?
- What policies exist to support rural service provision?
- How transferable are the models being explored and shared in this book to other locations outside the case countries?

Rural places and rural services

As noted, one of the critical elements to successful rural and small town communities is the availability of an appropriate suite of human services and accompanying service infrastructure. While 20th-century models of service delivery supported post-war rural and small town places, including the expansion of many resource-dependent places along the development fringes of a number of OECD states, the social, political, and economic restructuring that emerged in waves after the early 1980s disrupted those older models. In this section, we outline two issues of context. The first concerns the different ways by which rural and small town places are defined across the case studies. The second concerns a generalized model for understanding the transformations that have impacted rural service provision over time.

Defining rural and small town

Definitions of rural and small town places vary considerably in the literature and between national contexts. In Canada, definitions build on the work of the national statistical agency – Statistics Canada – which has developed a wide range of definitions of different types and levels of geographic groupings of populations (du Plessis *et al.*, 2004). Statistics Canada's 'census rural' definition refers to individuals living in the countryside outside centres of 1,000 or more population. In turn, the 'rural and small town' definition refers to individuals in towns or municipalities outside the commuting zone of larger urban centres (those with 10,000 or more population). This would include those enumerated under the census rural definition. These rural and small town locations may also be disaggregated into 'zones' according to the degree of influence of a larger urban centre (MIZ or 'metropolitan influenced zones').

In other jurisdictions, the 'numbers' assigned to definitional categories differ according to the uniqueness of each national settlement context. In many contexts, rural areas are simply the 'residual' areas and populations not captured by more sharply defined 'urban' areas. For example, Statistics New Zealand identifies two non-urban categories: 'rural centres' and 'other rural'. Of these two, only rural centres are defined. They are settlements 'with a population of 300 to 999 in a reasonably compact area that services surrounding rural areas' (Statistics New Zealand, 2017). Statistics New Zealand also incorporates measures of urban influence on those rural areas through a four-tier scale – high urban influence, moderate urban influence, low urban influence, and remote.

The Australian Bureau of Statistics (ABS) also starts its 'geography' of population data with a defined set of urban categories and leaves 'rural' as the residual (Australian Bureau of Statistics, 2016). The ABS identifies 'major urban' centres with populations of 100,000 or more (transitioning in name to 'significant urban area'), 'other urban' centres with populations between 1,000 and 99,999, 'bounded localities', and the 'rural balance', which represents the remainder of a state or territory. As part of recent changes, the classification 'urban centres and localities' is now in use and represents 'areas of concentrated urban development with populations of 200 people or more ... primarily identified using objective dwelling and population density criteria using data from the 2016 Census'.

In the UK, the separation of urban and rural begins with a much higher threshold – 10,000 people (UK Office of National Statistics, 2017). There are then six sub-categories of 'rural', including 'town and fringe', 'town and fringe in a sparse setting', 'village', 'village in a sparse setting', 'hamlets and isolated dwellings', and 'hamlets and isolated dwellings in a sparse setting'.

Beyond strict statistical interpretations of rural and small town, researchers have presented a variety of alternative definitional frameworks that include community characteristics and perceptions of identity. For example, Cloke (1977) describes a settlement continuum with 'rural' at one end and 'urban' at the other. Similarly, du Plessis *et al.* (2004) present the concept of 'degrees of rurality', which nicely accommodates various interpretations of rural and allows for community identity to be mixed in with numerical population, distance, or density thresholds.

Interest, whether from research, applied, or policy directions, in how rural and small town communities are transitioning under social, political, and economic change has led to a search for broader and more inclusive understandings of places. There is, for example, a need to incorporate both the spatial setting and social behaviours within that setting. This links well with efforts by researchers such as Cloke (1989, p. 173), who bridges the empirical and social definitions by suggesting that 'rural' involves: a) extensive land uses, b) small and generally low-order settlements, and c) a way of life which recognizes 'the environmental and behavioural qualities of living as part of an extensive landscape'.

From a policy perspective, definitions of rural and small town places carry political implications in the allocation of funding and responsibilities for services provision. Per-capita funding models or evaluation metrics especially affect decisions on a wide range of service delivery issues (Sullivan *et al.*, 2014). Under urban-centric or neoliberal policy frameworks, the unique context and needs of rural and small town places can be lost. More broadly, the discourse that serves to define conceptualizations of rural is essential for understanding and wrestling with the nuances of place. This matters because without a nuanced understanding of rural or small town contexts, the design of programs, policies, and funding frameworks can easily miss the mark. Instead, research increasingly highlights how supports for rural and small town services must be developed within a place-based framework.

Rural and small town services

Most developed nations have followed a similar pattern with respect to the provision of rural and small town services (Halseth and Ryser, 2006). Prior to World War II, individual rural places were more or less on their own aside from major state or private infrastructure such as roads and rail lines. This resulted in a great deal of unevenness in service delivery and availability. Wealthier rural areas were able to support a wider range and a better quality of services – something that had recursive benefits as better quality health and education services created better community futures.

In contrast, the 30-year post-WWII period was characterized by extensive welfare state service investments (in, for example, health, wellness, education, transportation, communication, and recreation), which greatly expanded the range of services available in rural communities and regions. This acted to level the service 'playing field', as these investments also came with the expansion of national standards for the many different types of services being delivered to all state taxpayers. It was under this framework that many new resource-dependent towns were created in resource frontier settings. The central challenge embodied by this period of investment, however, was that each individual service was delivered separately – usually through completely separate infrastructure, and under separate government ministry or agency jurisdiction.

Since the 1980s, we have witnessed social, political, and economic restructuring under a neoliberal framework (characterized by market-oriented, deregulatory, and non-interventionist government – see below). This has included an aggressive roll-back of public services. For rural areas, the roll-back has been especially problematic. With the loss, closure, or regionalization of services, there have been losses in local jobs, human capital and expertise, and the spending or use that supported the local economy and other local services. These service changes have impacted the very viability of rural places as those who need access to services are forced to leave. These 'eras', and the interlinkages between them, have had profound impacts on the range and quality of services and infrastructure now available in rural and small town places.

Theoretical foundations and development eras

The theoretical foundations for this collection build upon the critical transformations in industrial and public policy approaches that have impacted rural and small town areas since the 1950s. The immediate post-WWII era was marked by Fordist industrial and Keynesian public-policy approaches. After the early 1980s, extensive restructuring resulted from the shift to flexible-accumulation industrial models and neoliberal public-policy approaches. More recently, economic collapse and the unevenness of globalization have created periods of reactionary incoherence in both industrial models and public-policy approaches. Under the auspices of neoliberalization, for example, we also find many significant cases of neo-Keynesian market interventions by the state. While this period of reactionary incoherence creates many contradictions and challenges for rural development through temporary and/or ill-conceived policy and program interventions, it also creates opportunity, as the recent era is also marked by an openness to, and willingness for, countenance, innovation, and experimentation. This edited volume seeks to bring some consolidation to this period of change/opportunity and highlight themes of convergence and coherence around the future of rural service delivery.

In the following section, we briefly outline three eras of rural development in the post-WWII period to the present. The regime eras are situated within a Western industrialized historical setting. While there is tremendous variability between regions, these phases are well documented in the literature to capture macro industrial and ideological shifts which impact conditions in rural regions. For each era, we provide a brief description and highlight characteristics of rural regional investment.

Era 1: staples-based Keynesian

The adoption of a Keynesian public-policy approach, especially its economic stimulus component, coincided in the immediate WWII period with the need to address two imperatives. The first was employment, and how to re-employ the millions of soldiers returning from the war effort – a great many of whom just six or seven years earlier were the unemployed masses of the Great Depression. The second was how to address the massive infrastructure deficits that would be required to allow the wartime experience of industrial production and global supply-and-distribution chains to transition into the efficient post-war production of consumer goods.

In Canada, senior governments during this period followed a public-policy approach based on a model of industrial resource development (Williston and Keller, 1997). This led to a 25–30-year period of rapid economic and community growth across the rural regions in the country (Halseth *et al.*, 2004). High-quality local infrastructure was used to attract a stable workforce (and their families) to rural resource industry centres (Davis and Hutton, 1989; Horne and Penner, 1992). Similar policy actions during this era supporting the extension of industrial resource development into rural and hinterland regions are found across other OECD countries such as Australia (Argent, 2017), New Zealand (Connelly and Nel, 2017a; Nel, 2015), and Finland (Tykkyläinen *et al.*, 2017).

Overall, the era became a directed enterprise with senior government policy goals aimed at nation building and reconstruction. In Canada, resource endowments were imagined as a foundation for rural (and metropolitan) prosperity. Regional development strategies, such as the proliferation of growth pole strategies, became common programmatic responses to address regional disparities (Marchak, 2011; Savoie, 1992). As such, Canada's experience with substantial public investments during this period is commensurate with international trends, leading to the considerable expansion of rural regions and the establishment of foundations of service infrastructure and delivery aimed at standards parity (e.g., in health and education) across national space. These trends were followed in other OECD countries as well (Connelly and Nel, 2017b; Kotilainen *et al.*, 2017). Together, public policy and public/private sector investments led to economic growth in rural regions. From the 1950s to the 1970s, many such regions enjoyed development and prosperity. But the era also set the stage for increased dependence on natural resource sectors and increased vulnerability to change in the global economy (Freudenburg, 1992).

Era 2: neoliberalism

The global recession of the early 1980s marked a considerable shift in the investment orientation and ideology of senior governments in favour of neoliberalism, with vast implications for resource economies and rural regions. Harvey (2005, p. 2) describes neoliberalism as "a theory of political economic practices that proposes that human well-being can best be advanced by liberating individual entrepreneurial freedoms and skills within an institutional framework characterized by strong private property rights, free markets, and free trade". The neoliberal response has expressed itself in a variety of economic and policy terms.

Economically, many rural regions within developed economies are high-cost commodity producers. They are high cost for many reasons: wages, taxation, regulation, services, etc., relative to some other regions of the globe. The process of globalization has rapidly opened up markets to the entry of resource commodities from low-cost producer regions (see Halseth and Ryser, 2018). The government response, faced with this international competition and enhanced industrial flexibility, has been to reduce the cost burden on resource industries in a variety of ways, via the liberalization of property rights, the liberalization of market regulations through the reduction of barriers for increased mobility of resource companies, and the liberalization of spaces by increasing access to natural resources (Heisler and Markey, 2013; Hreinsson, 2007; Tonts and Haslam-McKenzie, 2005; Young and Matthews, 2007). The central goal of senior levels of government in these roll-back strategies was to promote a stable and incentivized jurisdictional environment for resource industries.

The investment orientation of senior governments during this period may be best characterized as state withdrawal. This social policy response is characteristic of a shift in government policy from an equity-based orientation to less defined attempts at enabling regional and community development (Polèse, 1999). This means that successive governments have been gradually withdrawing from a commitment to provide equitable access to standardized services across rural and urban space, while making modest (and incomplete) efforts to assume a secondary role of facilitating transition through various community and regional development programs (Markey *et al.*, 2007). This change in approach is partly driven by demands for greater bottom-up representation and control; however, many of the key fiscal and policy levers that communities and rural regions need to mobilize and realize those strategies are still held firmly by the state. Without the support of top-down public policy, local and regional initiatives struggle, leading to the downsizing or closure of rural and small town services, and the gradual decay of infrastructure investments from the previous era (Sullivan *et al.*, 2014). While many of the service delivery models of the 1950s and 1960s are no longer practical in regions with dispersed sets of small communities, rather than seeking to deploy new models that make use of innovations in organization and technology, states too often simply close or regionalize services outright using the neoliberal-inspired metric that costs of delivery are too high when measured against urban examples.

A critical feature of neoliberal restructuring for rural and small town places is that the state has not yet been able to forge a coherent response to the post-1970s/1980s restructuring of the political economy of rural regions (Woods, 2007). Their policy mindset has changed and investments have been reduced, but their tools (the structure of fiscal and policy power and authority) for supporting rural and small town places have not changed. This has led to the current regime of incoherence as the pressure builds to replace or renew critical infrastructure and services. This lag in investment has supported the well-documented infrastructure 'deficits' that exist across the rural areas of many OECD states. The failure to adopt new, and better suited, service delivery models and to attend to critical infrastructure deficits impacts the capacity of rural and small town places to address both economic and social reproduction. It also limits their resilience and the support mechanisms required to mitigate the negative impacts associated with resource sector decline/closure and to engage with the opportunities simultaneously being created in the global economy.

Era 3: reactionary incoherence

There are a number of competing and divergent trends currently expressing themselves within rural resource regions. The actions and outcomes of various actors are too chaotic and idiosyncratic to each circumstance or jurisdiction to defy the kind of logic or coherence implied by the terms 'model' or 'framework' that more neatly captures the depth and durability of staples theory and neoliberalism. Instead, we label this new policy era as 'reactionary incoherence'. This label incorporates a sense of the chaotic and disorganized elements that support a level of policy incoherence. It also suggests a mix of retrenchment and opportunistic initiatives/reactions undertaken specifically to maintain a past hegemonic structure and set of actors and relationships.

There are a variety of dynamic factors at work driving this era. First, there are the many inherent contradictions within neoliberalism itself (Halseth and Ryser, 2018; Harvey, 2005). This includes the rush right after the global economic collapse of 2008 by numerous state governments (regardless of political affiliation) to deploy massive state spending and other interventions in the economy to support employment, encourage production and consumption, and generally provide stimulus during times of economic collapse. While this approach has been effective at staving off several short-term impacts, particularly in rural regions (e.g., employment opportunities associated with long-awaited infrastructure investments), the investments, reactionary in nature, are hindered by a lack of regional knowledge and vision (i.e., reactionary investments based on 'shovel-ready' projects, rather than a coherent approach to establishing the appropriate infrastructure for supporting transformative rural change in the 21st century). Senior governments have reduced rural and regional offices, leaving centralized bureaucracies with a lack of 'eyes and ears on the ground'. This physical disengagement, decades of relative indifference to rural investments, and lack of a coherent new vision for rural development all mean that state response to dramatic rural needs too often has been to retreat to older economic models, losing opportunities associated with what could have been strategic investments for the future (Kim and Warner, 2018; Morris, 2015).

Second, the 'end of life cycle' status of much of the critical infrastructure in rural regions (built during the immediate post-WWII decades but run down by three decades of neoliberal policy) is gaining more attention (Argent, 2013; CCC, 2013; FCM, 2012). Degraded infrastructure impacts communities in numerous ways. There may be severe health implications associated with services like drinking water. From an economic standpoint, failing or inadequate infrastructure impedes new economic opportunities, makes existing activities more expensive (and therefore less competitive), and reduces the likelihood that communities will be able to attract and retain both people and capital. For our purpose, the removal or degradation of rural services decreases livability, hinders economic competitiveness, negatively impacts community development capacity, and diminishes the well-being of rural residents.

Senior governments have been offloading the responsibility for infrastructure to the local level, meaning municipalities and localities are now responsible for a greater share of infrastructure. In Canada, for example, it is local governments that now own more than 60% of all local infrastructure, yet those same local governments have the fewest fiscal tools and least fiscal capacity to address mounting costs when compared to other levels of government in the country

(The Canadian Chamber of Commerce (CCC), 2013; Federation of Canadian Municipalities, 2012; Fletcher and McArthur, 2010). This presents a tremendous challenge to rural communities where low population levels, low densities, large distances, and more extreme weather conditions increase the per capita cost of critical infrastructure (CCME, 2006; CRRF, 2015; Rolfe and Kinnear, 2013). Overall, the recognized and well-researched need for infrastructure renewal is constrained both in governance and implementation efficacy, leading again to policy uncertainty.

Third, there is considerable nostalgia for the Keynesian investment era. This should not be conflated, however, with a desire to return to such a passive role for communities and regions regardless of the investment dollars. Stressed by decades of neglect and emboldened by the importance of community and regionalist approaches to development, rural communities are now active players in their own development (Makuwira, 2007; Manson *et al.*, 2016; Sørensen, 2017). Co-constructed development pathways (i.e., those that involve the top-down and bottom-up) have been proven to be very effective (Cheshire *et al.*, 2014; Shucksmith, 2010). The opportunity to leverage senior government fiscal capacity and regulatory control with highly contextualized knowledge leads to better development outcomes. However, communities and regions are limited by a lack of institutional capacity and the mechanisms to sort priorities and ongoing responsibilities remain weak. Senior governments also remain loath to transfer or relinquish regulatory or fiscal control. As Halseth states:

> While senior government has instructed communities to be more entrepreneurial in searching for economic opportunities and attracting new business ventures, it has at the same time removed many of the critical supports necessary to help communities secure those new economic activities or businesses.
>
> (2017, p. 5)

Finally, adding to the complexity is the uncertain role of industry in rural resource regions (Cheshire, 2010). During the period of neoliberalism, industries shifted dramatically away from their historic paternal relationship with resource-dependent communities forged during the staples era. Given the concomitant withdrawal of senior government roles in terms of representing and safeguarding the public interest relative to resource sectors, industry now finds itself having to rebuild community and regional relationships under the guise of corporate social responsibility, impact-benefit agreements, and other mechanisms. Resource companies now must contend with community and regional expectations, developed in response to decades of neglect, that corporate initiatives obtain a form of local 'social license' for new resource industry projects (Heisler and Markey, 2014). The challenge with these new mechanisms is that they are institutionally weak, are not regulated, and lack oversight and monitoring. All of which again leads to confusion and policy incoherence.

All of these factors serve as both outcomes and indicators of broader policy incoherence and drivers of ongoing uncertainty. Ultimately, they all contribute to questions about the most appropriate ways and mechanisms for re-investing in rural places, and about the appropriate roles for senior and local governments in terms of directing development and representing the broader public interest relative to resource economies. There has been a growing and recognized need to re-invest in rural regions, the challenge being uncertainty about how, how much, where, and for what.

Book organization

This book is organized into five parts. Starting with our introduction, Greg Halseth, Laura Ryser, and Sean Markey briefly discuss the critical role of services and infrastructure in rural and small town places. The chapter highlights how older models of service delivery, and their supporting infrastructure, are increasingly not suited to a 21st-century context. As so often happened through the mid-20th century, individual services and the supporting infrastructure of buildings were created as single-purpose entities to deliver only that service in isolation from other services and community needs. Renewing and replacing outdated single-purpose infrastructure and services in small places is significantly challenged by the fiscal and policy realities of the 21st century. The book addresses this challenge with research from Australia, Canada, Ireland, New Zealand, and the UK. The chapters focus on service and infrastructure models capable of strengthening rural and small town resiliency using three parts of discussion: government policies, new governance and funding arrangements to support new service models, and alternative infrastructure arrangements.

Beginning with the important theme of shaping new service arrangements through government policies, Chapters 2–4 explore local and senior government policy approaches that have supported new service and infrastructure models. In this part, Erin Sherry and Sally Shortall describe the rural proofing concept used in Northern Ireland to allow policy makers to assess the equitable treatment of rural communities based on their service needs and circumstances. Based on the Rural Needs Act (2016), government departments, local councils, and other public bodies are required to consider rural needs in all strategies, policies, and plans related to the delivery of services. Their research critically reviews the assumptions underpinning rural proofing and rural need, and questions whether the legislative approach as pioneered is more likely to help or hinder innovative service delivery.

In Chapter 3, Greg Blackburn describes recent efforts in Tasmania, Australia, to reform major public sector services through Service Tasmania, the first whole-of-government initiative. The project was designed to improve service delivery for those living in rural and remote areas that were previously constrained by fragmented and inefficient bureaucratic processes. By adopting a single integrated one-stop approach to providing seamless, cross-agency government service delivery, Service Tasmania offers almost 600 government services to the community through physical, online, and telephone channels. This chapter reflects on the lessons learned through Australia's first whole-of-government initiative. The chapter profiles the benefits rural and urban communities obtained through this customer-orientated shift in government service delivery, the catalyst for change, and implementation challenges associated with the approach.

In the third chapter of Part II, Brian Dollery offers insights into shared service models used by local governments in Australia. Small regional, rural, and remote local authorities often face numerous constraints, including difficulties associated with attracting staff with specialist administrative and technical skills. Shared services represent a method for overcoming these difficulties, and web-based systems can facilitate the delivery of these services in an efficient fashion over long distances. In Australian local government, this mode of shared service provision has been pioneered by the Brighton Council in the form of a common service provision model. In contrast to most existing shared service platforms in Australian local government, the common service provision model is wholly owned by the Brighton Council, and it provides the same council functions and services to other local government entities on a commercial 'fee-for-service' basis. This chapter considers the national and state policy parameters within which local authorities must operate as they pursue these models, and offers a brief assessment of the policy implications to support these approaches in small communities.

Part III includes four chapters that highlight new funding and governance arrangements to enhance the resiliency of small town services, with a particular focus on health care services. In Chapter 5, Etienne Nel and Sean Connelly begin by presenting some of the rural service delivery challenges that have unfolded in New Zealand in an era of neoliberalism and new managerialism. The introduction of new managerialism practices has passed particular service delivery responsibilities directly on to rural communities, which are often under-resourced to take on these new responsibilities. These shifts have resulted in greater reliance on the social infrastructure (e.g., volunteers, networks, and community groups) in communities to adapt and respond to economic and social change. Drawing on evidence from the towns of Lawrence, Clyde, and Tapanui, located on the South Island, they explore how communities are responding, the opportunities and limitations of new models, and the implications for future service delivery in rural places. In more resilient communities, rural health trusts have been formed and the community has taken over the operations of health care facilities. However, this is not the case in communities where social capital and resourcing levels are weaker, leaving them at a structural disadvantage which reinforces polarizing spatial and social differences across the rural areas of the country.

In Chapter 6, Neil Hanlon, Martha MacLeod, Trish Reay, and David Snadden explore partnerships for health care sustainability in smaller urban centres in northern British Columbia, Canada. Health care systems that serve predominantly rural and remote populations face a number of challenges (e.g., health professional recruitment and retention difficulties, gaps in service, poor service integration, and continuity) that require a concerted set of policy responses. In Canada, several national and provincial health commissions have recommended a bundling of patient-focused health reforms, health human resource strategies, and primary health care development as the means to resolve longstanding rural health care shortcomings. Yet such comprehensive policy efforts have remained elusive. Reforms promising primary health care development and upstream population health approaches have not advanced past the stage of small, limited-term pilot initiatives. A notable exception is a primary health care reform initiative currently underway in northern BC, Canada, where partnerships with community leaders have been developed to achieve support for wellness promotion campaigns and community-based primary health care reform.

Chapter 7 builds upon the book's examination of rural health care. Rachel Winterton, Kaye Knight, Catherine Morley, and Wendy Walters examine community and residential models of dementia care in rural and small town communities. Australian neoliberal policy discourses around health and aged care strongly advocate for locally sensitive responses to local health care needs, while simultaneously undertaking continued service rationalization and centralization. This places significant pressure on small, remote health care services to develop and implement diverse local models of care, in conjunction with other local partners and community members. Their case profiles the ABLE-D dementia model of care developed by one small rural health service in northwest Victoria, Australia. This program sought to tailor local community and residential dementia services to meet the needs of their communities, while taking into account the administrative, resource, and geographic constraints of the organization. In discussing factors associated with the model's sustainability after five years, the authors highlight some of the benefits, lessons, and challenges associated with implementing and sustaining this model.

In Chapter 8, Ryan Gibson and Joshua Barrett complete this section by exploring alternative funding options offered by philanthropic organizations to strengthen the sustainability of rural services and infrastructure. They profile two rural community foundations: Virden Area Foundation (Manitoba) and the Sussex Area Community Foundation (New Brunswick) in Canada. The transition to alternative funding services has been facilitated via the withdrawal of public investment, the retreat of service delivery, and the discontinuance of non-profit organizations. Although neither philanthropic organization intentionally sought to provide service delivery, they have recognized the critical importance of these services in building sustainable communities.

In a context of aging infrastructure, high operating costs, and budget constraints, the fourth part of the book consists of three chapters selected to portray the different ways in which rural stakeholders are pursuing new infrastructure arrangements. In Chapter 9, Sarah Minnes, Sarah-Patricia Breen, and Kelly Vodden examine some of the innovations for sustaining rural drinking water services in Canada. Since the 1980s, a prolonged period of economic, social, and political restructuring has resulted in the offloading of responsibilities from senior to local governments, and decreased levels of investment. This history, combined with recent economic, political, social, and environmental changes; changing regulatory environments; a growing infrastructure deficit; and increased expectations for service delivery, continues to impact rural communities and regions across the country. This research demonstrates innovative, place-based solutions in tackling the infrastructure deficit and other challenges through training and asset management activities at community and regional-scale levels based on case studies in Newfoundland and Labrador and British Columbia.

In Chapter 10, Wayne Kelly and Michael Hynes examine proposals and strategies for the development of rural broadband in two countries – Canada and Ireland – providing an overview of pressures and barriers to the roll-out of telecommunications infrastructure to rural economies when market forces are reluctant or failing to do so. Policies for rural development have traditionally centred on the exploitation of land-intensive natural resources such as agriculture and forestry, but more contemporary approaches aim to contribute to recognizing and making use of strengths and opportunities that are linked to locally produced economic development strategies. This has led to debates on the role the telecommunications sector should play as a key contributor to (re)developing rural societies and economies. The chapter highlights how many developed countries struggle with the challenge of extending their broadband infrastructure to rural and remote areas.

Policies that are driving new expectations for integrated or shared service arrangements are challenging the transformative capacity of rural organizations. At the same time, small communities are confronted with the challenges of aging and inadequate infrastructure established in the post-WWII era. In Chapter 11, Laura Ryser, Greg Halseth, and Sean Markey investigate this theme through an examination of 16 case studies associated with co-location infrastructure arrangements in British Columbia, Canada. A key issue identified in the chapter is that there is no central hub for rural stakeholders to learn about different models and processes that have been used to develop and operate these assets. There is also a limited understanding of ownership and user agreements, design features that can improve the functionality of multi-purpose spaces, issues of risk and liability, and protocols to guide the development, operations, and maintenance of these facilities. Given the limited tax base of many small communities, the chapter highlights how greater flexibility is needed to support financing arrangements and planning for these complex rural infrastructure initiatives.

In the final part of the volume, we revisit the strategic policies, planning, and investments needed to support alternative service and infrastructure models, while also highlighting some ongoing challenges that shape resiliency across different rural settings. The capacity of stakeholders to work with local and non-local partners to strategically develop appropriate new models is contingent on acknowledging the significance of these communities in our broader regional and global economies. In this final chapter, we revisit key emerging issues in both policies and community assets that shape the resilience and renewal of rural and small town places through more integrated service and infrastructure arrangements. The chapter also details a research agenda to advance rural community development research. We conclude with reflections on broader policy, practice, and theoretical implications for more integrated service and infrastructure models for 21st-century rural and small town sustainability.

It is our hope that this book will be of value to a wide variety of government, policy, research, private sector, and community audiences interested in rural and small town communities and regions, and social and economic restructuring within those places, as well as those interested in the more general topics of community and regional development.

References

Argent, N. 2013. "Reinterpreting core and periphery in Australia's mineral and energy resources boom: An Innisian perspective on the Pilbara", *Australian Geographer* 44(3): 323–340.

Argent, N. 2017. "Trap or opportunity? Natural resource dependence, scale, and the evolution of new economies in the space/time of New South Wales' Northern Tablelands". In G. Halseth (ed.), *Transformation of resource towns and peripheries* (pp. 18–50). London: Routledge.

Australian Bureau of Statistics. 2016. *Australian Statistical Geography Standard (ASGS): Volume 4: Significant urban areas, urban centres and localities, section of state, July 2016.* Available online at www.abs.gov.au/ausstats/abs@.nsf/mf/1270.0.55.004. Accessed 20 December 2017.

Brown, D.L. and Schafft, K.A. 2011. *Rural people and communities in the 21st century: Resilience and transformation.* Cambridge: Polity Press.

The Canadian Chamber of Commerce (CCC). 2013. *The foundations of a competitive Canada: The need for strategic infrastructure investment.* Ottawa: The Canadian Chamber of Commerce.

Canadian Council of Ministers of the Environment. 2006. *Examination of potential funding mechanisms for municipal wastewater effluent (MWWE) projects in Canada.* Winnipeg: Canadian Council of Ministers of the Environment.

Canadian Rural Revitalization Foundation (CRRF). 2015. *State of rural Canada report.* Available online at: sorc.crrf.ca. Accessed 13 February 2018.

Cheshire, L. 2010. "A corporate responsibility? The constitution of fly-in, fly-out mining companies as governance partners in remote, mine-affected localities", *Journal of Rural Studies* 26(1): 12–20.

Cheshire, L., Everingham, J.A., and Lawrence, G. 2014. "Governing the impacts of mining and the impacts of mining governance: Challenges for rural and regional local governments in Australia", *Journal of Rural Studies* 36: 330–339.

Cloke, P. 1989. "Rural geography and political economy". In R. Peet and N. Thrift (eds.), *New models in geography: The political-economy perspective* (Vol. 1, pp. 164–197). London: Unwin Hyman Ltd.

Cloke, P.J. 1977. "An index of rurality for England and Wales", *Regional Studies* 11(1): 31–46.

Connelly, S. and Nel, E. 2017a. "Restructuring of the New Zealand economy: Global-local links and evidence from the West Coast and Southland regions". In G. Halseth (ed.), *Transformation of resource towns and peripheries: Political economy perspectives* (pp. 112–135). Abingdon: Routledge.

Connelly, S. and Nel, E. 2017b. "Community responses to restructuring". In G. Halseth (ed.), *Transformation of resource towns and peripheries: Political economy perspectives* (pp. 317–335). Abingdon: Routledge.

Davis, H. and Hutton, T. 1989. "The two economies of British Columbia", *BC Studies* 82: 3–15.

du Plessis, V., Beshiri, R., Bollman, R., and Clemenson, H. 2004. "Definitions of rural". In G.Halseth and R. Halseth (eds.), *Building for success: Explorations of rural community and rural development* (pp. 51–80). Brandon, MB: Rural Development Institute.

Federation of Canadian Municipalities. 2012. *Canadian infrastructure report card Volume 1: 2012 municipal roads and water systems.* Ottawa: Federation of Canadian Municipalities. Available online at www.fcm.ca/Documents/reports/Canadian_Infrastructure_Report_Card_EN.pdf. Accessed 14 February 2018.

Fletcher, J. and McArthur, D. 2010. *Local prosperity: Options for municipal revenue growth in British Columbia.* Vancouver: Think City.

Freudenburg, W. 1992. "Addictive economies: Extractive industries and vulnerable localities in a changing world economy", *Rural Sociology* 57(3): 305–332.

Halseth, G. 2017. "Introduction: Political economy perspectives on the transformation of resource towns and peripheries." In G. Halseth (ed.), *Transformation of resource towns and peripheries: Political economy perspectives* (pp. 1–10). Abingdon: Routledge.

Halseth, G., Markey, S., and Bruce, D. (eds.). 2010. *The next rural economies: Constructing rural place in a global economy.* Oxfordshire, UK: CABI International.

Halseth, G. and Ryser, L. 2006. "Trends in service delivery: Examples from rural and small town Canada, 1998 to 2005", *Journal of Rural and Community Development* 1(2): 69–90.

Halseth, G. and Ryser, L. 2018. *Towards a political economy of resource dependent regions.* Abingdon: Routledge.

Halseth, G., Straussfogel, D., Parsons, S., and Wishart, A. 2004. "Regional economic shifts in British Columbia: Speculation from recent demographic evidence", *Canadian Journal of Regional Science* 27(3): 317–352.

Harvey, D. 2005. *A brief history of neoliberalism.* Oxford: Oxford University Press.

Heisler, K. and Markey, S. 2013. "Scales of benefit: Political leverage in the negotiation of corporate social responsibility in mineral exploration and mining in rural British Columbia, Canada", *Society & Natural Resources* 26(4): 386–401.

Heisler, K. and Markey, S. 2014. "Navigating jurisdiction: Local and regional strategies to access economic benefits from mineral development", *The Canadian Geographer* 58(4): 457–468.

Horne, G. and Penner, C. 1992. *British Columbia community employment dependencies.* Victoria, BC: Planning & Statistics Division and Ministry of Finance & Corporate Relations.

Hreinsson, E. 2007. "Deregulation, environmental and planning policy in the Icelandic renewable energy system". In *2007 international conference on clean electrical power.* Capri, Italy. May, pp. 283–290.

Kim, Y. and Warner, M.E. 2018. "Geographies of local government stress after the great recession", *Social Policy & Administration* 52(1): 365–386.

Kotilainen, J., Halonen, M., Vatanen, E., and Tykkyläinen, M. 2017. "Resource town transitions in Finland: Local impacts and policy responses in Lieksa". In G. Halseth (ed.), *Transformation of resource towns and peripheries: Political economy perspectives* (pp. 296–316). Abingdon: Routledge.

Makuwira, J. 2007. "The politics of community capacity-building: Contestations, contradictions, tensions and ambivalences in the discourse in Indigenous communities in Australia", *The Australian Journal of Indigenous Education* 36(S1): 129–136.

Manson, D., Markey, S., Ryser, L., and Halseth, G. 2016. "Recession response: Cyclical problems and local solutions in northern British Columbia", *Tijdschrift voor Economische en Sociale Geografie* 107(1): 100–114.

Marchak, P. 2011. *Green gold: The forest industry in British Columbia.* Vancouver: UBC Press.

Markey, S., Halseth, G., and Manson, D. 2007. "The (dis)connected north: Persistent regionalism in northern British Columbia", *Canadian Journal of Regional Science* 30(1): 57–78.

Markey, S., Halseth, G., and Manson, D. 2012. *Investing in place: Economic renewal in northern British Columbia.* Vancouver: UBC Press.

Morris, G. 2015. *Who cares about rural England's disadvantaged now? The implications of the closure of the Commission for Rural Communities for the disadvantaged people and places of rural England.* Exeter: University of Exeter. Available online at https://ore. exeter.ac.uk/repository/bitstream/handle/10871/17824/Who%20cares%20about%20dis advantage%20now%2012%20February%202015.pdf?sequence=1&isAllowed=y.

Nel, E. 2015. "Evolving regional and local economic development in New Zealand", *Local Economy* 30(1): 67–77.

OECD. 2010. *Strategies to improve rural service delivery.* Paris: OECD.

OECD. 2014. *Innovation and modernising the rural economy.* Paris: OECD.

Polèse, M. 1999. "From regional development to local development: On the life, death, and rebirth(?) of regional science as a policy relevant science", *Canadian Journal of Regional Science* 22(3): 299–314.

Rolfe, J. and Kinnear, S. 2013. "Populating regional Australia: What are the impacts of non-resident labour force practices on demographic growth in resource regions?", *Rural Society* 22(2): 125–137.

Savoie, D. 1992. *Regional economic development: Canada's search for solutions.* Toronto: University of Toronto Press.

Shucksmith, M. 2010. "Disintegrated rural development? Neo-endogenous rural development, planning and place-shaping in diffused power contexts", *Sociologia Ruralis* 50(1): 1–14.

Sørensen, T. 2017. "Community development in an age of mounting uncertainty: Armidale, Australia". In G. Halseth (ed.), *Transformation of resource towns and peripheries: Political economy perspectives* (pp. 249–267). Abingdon: Routledge.

Statistics New Zealand. 2017. *Defining urban and rural New Zealand.* Available online at http://archive.stats.govt.nz/browse_for_stats/Maps_and_geography/Geographic-areas/ urban-rural-profile/defining-urban-rural-nz.aspx. Accessed 19 December 2017.

Sullivan, L., Ryser, L., and Halseth, G. 2014. "Recognizing change, recognizing rural: The new rural economy and towards a new model of rural service", *The Journal of Rural and Community Development* 9(4): 219–245.

Tonts, M. and Haslam-McKenzie, F. 2005. "Neoliberalism and changing regional policy in Australia", *International Planning Studies* 10(3–4): 183–200.

Tykkyläinen, M., Vatanen, E., Halonen, M. and Kotilainen, J. 2017. "Global – Local links and industrial restructuring in a resource town in Finland: The case of Lieksa". In G. Halseth (ed.), *Transformation of resource towns and peripheries: Political economy perspectives* (pp. 85–111). Abingdon: Routledge.

UK Office of National Statistics. 2017. *The 2011 rural-urban classification for small area geographies: A user guide and frequently asked questions.* Available online at www.gov. uk/government/uploads/system/uploads/attachment_data/file/239478/RUC11user_gui de_28_Aug.pdf. Accessed 20 December 2017.

Williston, E. and Keller, B. 1997. *Forests, power, and policy: The legacy of Ray Williston.* Prince George, BC: Caitlin Press.

Woods, M. 2007. "Engaging the global countryside: Globalization, hybridity and the reconstitution of rural place", *Progress in Human Geography* 31(4): 485–507.

Young, N. and Matthews, R. 2007. "Resource economies and neoliberal experimentation: The reform of industry and community in rural British Columbia", *Area* 39(2): 38–51.

Part II

Shaping new service arrangements through government policies

2 The needy rural

Does living in a rural area mean that you are in need?

Erin Sherry and Sally Shortall

Introduction

This chapter considers rural proofing, a form of rural mainstreaming, as a method of measuring whether access to services in rural areas is problematic or success-ful. Similar to gender mainstreaming, rural mainstreaming does not identify a particular problem but says all policies must ensure they do not discriminate against rural areas, particularly in relation to service provision. Rural main-streaming is a conceptual tool intended to deliver fair and equitable treatment by ensuring that the particular needs and circumstances of rural areas are routinely considered across government. In practice, rural mainstreaming can be complex and take on various and multiple forms and characterizations: for example, in England and Northern Ireland, key phrases are *rural proofing* and *rural champion;* in Canada it is *rural lens;* and most recently in Northern Ireland it is *rural needs*. The *rural proofing* and *rural champion* models are relatively unique to England (Organization for Economic and Co-operative Development [OECD], 2011), with similar approaches appearing in some parts of the United Kingdom (UK) and other countries with historic or institutional ties such as Sweden, Canada, and New Zealand. Rural proofing is the means of ensuring all policies are evaluated to see if they have an adverse impact on rural areas. The rural champion is the body tasked with ensuring that rural proofing takes place, and in England and Northern Ireland they are the departments with responsibi-lities for rural affairs. In this chapter, we are focusing on the recent legislation to address rural needs in Northern Ireland and what impact it will have on service provision.

Rural mainstreaming models emerged largely out of the increasing awareness and resourcing of rural development policy and initiatives as the structure of rural economies diversified. They differ from traditional service delivery models in that the idea is not to develop specific rural policy but rather to review existing and new policies to ensure that urban and rural residents receive equitable access. While it is recognized that there are some distinctive aspects to the delivery of public resources in rural areas, the approach hinges on the assumption that these issues can be addressed during the design and development of general policies (OECD, 2011). It assumes that rural and urban are not that different. It is not that rural areas need specifically designed policies to meet their particular issues, rather policies can be tweaked to meet rural need.

The evolution of rural mainstreaming in Northern Ireland from the English model of 'rural proofing' to legislating[1] explicitly for 'rural need' is considered in this chapter. The analysis is based on qualitative research carried out in Northern Ireland on the previous iterations of proofing commitments, as well as issues stemming from the new needs-based initiative. This chapter critically reflects on the assumptions underpinning 'rural proofing' and 'rural need', and questions whether the legislative approach as pioneered is likely to generate any meaningful benefits. It is argued that while in theory 'rural proofing' should address the delivery of all services in rural areas, the reality is that it does not. In practice, it is vague and emotive, which renders it an ineffective policy. It is also argued that no evidence of particular rural service provision needs has been provided or used to underpin the Rural Needs Act.

The chapter is structured as follows: The next section provides background on rural proofing more generally. This is followed by context specific to the study area. The key findings from analysis of existing policy documents and primary qualitative data are then described, and finally, the ability of a novel 'rural proofing' model, based on a due regard to 'rural needs', to ensure service delivery is discussed.

Rural proofing and rural champion

The European Union re-evaluated Common Agricultural Policy (CAP) in the 1980s due to the fact that it had fostered a market distorting agricultural industry while rural areas declined (Bryden, 2009; Bryden *et al.*, 2010; Copus *et al.*, 2006; Shortall, 1996; Shortall and Alston, 2016; Shucksmith *et al.*, 2005). The CAP was expanded to include a rural development program: an area-based approach designed to consider social, cultural, and economic activities unrelated to agriculture (Shortall and Alston, 2016). Numerous studies have considered the benefits and drawbacks of this policy shift (Bock, 2004; Shortall and Shucksmith, 2001; *Sociologia Ruralis*, 2000). An unanticipated consequence was that European member states began to scrutinize national policies and to consider rural policy more generally (Shortall, 2012, 2013).

Mainstreaming is a policy based on reviewing government activities to ensure that all groups receive comparable treatment. The OECD report (2011) describes the English rural proofing process as 'rural mainstreaming', presumably to make it understandable to international audiences. England is unique among OECD countries in the way in which it has developed rural mainstreaming around 'rural proofing' and 'rural champion' (Atterton, 2008; OECD, 2011; Shortall and Alston, 2016). The idea is not to develop specific rural initiatives but rather to check that urban and rural residents receive equitable access to a common set of government interventions. The 'rural proofing' model is based on the premise that, while rural areas may exhibit unique characteristics, including some element of disadvantage, accommodation may be met within universal government design and development processes (see OECD, 2011). In England's first iteration of rural proofing, an independent body was the 'rural champion' or 'rural watchdog'. More recently, this function has been assumed by the government department with responsibility for rural affairs.

Some Commonwealth and European countries have considered adopting similar models. However, the dominant examples remain within the UK: England and Northern Ireland. Until recently, Northern Ireland had almost entirely mirrored the English model (Sherry and Shortall, Forthcoming). However, Northern Ireland has now gone further than any other country or region by passing into legislation the Rural Needs Act (Northern Ireland) 2016. The act is structured around the premise that the way to address rural needs is through placing duty of due regard, or statutory rural proofing of government strategies, policies, and plans, as well as the delivery of services (Department of Agriculture, Environment and Rural Affairs [DAERA], 2017).

While it is logical to expect that pressures around rural service provision and rural need led to the introduction of the Rural Needs Act, this is not actually the case. The guide to the Rural Needs Act (2018) reports on distance to services, with education and health care used as the two services deemed to be human rights (Jack *et al.*, 2012). In Northern Ireland, settlements of 10,000 or more are deemed to offer full services. Using this approach, areas have been classified as being either 'within' or 'outside' 20-minute or 30-minute drive times of their nearest town centre. Approximately 93% of Northern Ireland's population lives within 30 minutes' drive time of the town centre of a settlement containing a population of at least 10,000. 80% live within 20 minutes' drive time. The guide to the Rural Needs Act does not reference one single example of need or pressure for rural services. It has also been argued that the political structure in Northern Ireland is such that until the devolved government in 2002, civil servants simply implemented UK policy. The political context is one where there is still a certain nervousness and reluctance to assume responsibility for designing and executing policy. There is little tradition of engaging with expertise beyond the civil service (Knox, 2008; Shortall, 2012). Knox (2008) has written a scathing article about the review of public administration in the region and titled his paper 'Ignoring the Evidence'. The same lack of relevant evidence to underpin the development of the Rural Needs Act is evident here. There is also evidence of the continuing tendency to copy English policies.

Rural proofing has historically relied on training, procedures, and auditing (see for example, Department of Agriculture and Rural Development [DARD], 2015; Department for Environment, Food and Rural Affairs [DEFRA], 2015) instead of the identification of specific goals, targets, or outcomes (Shortall and Alston, 2016). The drawbacks of applying a single checklist to deal with the diverse issues faced by rural areas have been raised – particularly the absence of a rural voice, as communities are not directly consulted (Atterton, 2008). It has also been argued that problematic feedback loops can arise, whereby insufficient theoretical groundwork and inherent methodological weaknesses reinforce one another in the development of rural policy (Sherry and Shortall, Forthcoming).

Problems with the practical implementation of 'rural proofing' have been identified, including inconsistent application across government departments, assignment of responsibility to junior staff, decreasing visibility and awareness, weak leadership, and ineffective monitoring (OECD, 2011). England found that over half of impact assessments across government failed to consider rural issues despite there being a likely impact: over one-third discussed rural issues, but not any policy implications, and only one-tenth were considered to contain robust evidence on 'rural proofing' or any role in terms of policy design (Department for Environment, Food and Rural Affairs, 2015). Similarly, disappointment was observed in Northern Ireland, particularly with the ability for 'rural proofing' to influence policy making, most likely linked to mechanical difficulties in making use of the approach (DARD, 2015). The context and history of rural proofing leading to legislation in Northern Ireland is now considered.

Rural proofing in Northern Ireland

Northern Ireland has a land area of just over 5,000 square miles (13,000 square kilometers) and a population estimated at just over 1.8 million in June 2015 (Northern Ireland Statistics and Research Agency [NISRA], 2017). The capital city and main urban settlement within Northern Ireland is Belfast, situated in the East. Belfast is about 110 kilometers (70 miles) from the second city, Londonderry, located in the West, and about 160 kilometers (100 miles) from Dublin to the South, the capital of Ireland. Historically, social and economic factors have contributed to an East/West pattern of relative affluence/deprivation that overrides a strict urban/rural dichotomy. For example, the largest share of employment income considering the population of working adults is in the NUTS 3 region of Outer Belfast, while the lowest relative share is earned in the West and the South (Sherry and Wu, 2018). When the statistical regions correspond to the national designations of rural East and rural West, the pattern persists, as these two rural areas exhibit the lowest and highest rates of individuals in relative poverty respectively, with the remaining three urban designations falling somewhere between them (Department for Social Development [DSD], 2016). In other words, the urban/rural distinction is not the most significant one in Northern Ireland.

Official designations of rural, such as the criteria for determining eligibility for European Union funding for rural development, are dichotomous (urban-rural), with a distance threshold but no socioeconomic dimension. An inter-departmental working group established a standard urban-rural definition, particularly for use in relation to European Union funding requirements, designating settlements with a population of below 4,500 as rural, resulting in 35% of the population classified as living within a rural area (NISRA, 2005). This definition was reviewed in 2015 in light of a new census made available and encouragement from the Northern Ireland DARD as part of its Rural White Paper Action Plan (NISRA, 2015). The revised recommendations reflect an adjustment in the definition of an intermediate settlement, the boundary between urban and rural, increasing to a population of 5,000 within a settlement boundary.

The review also introduces estimated drive times to a service centre (a settlement with a population of 10,000 or more) to further distinguish between 'accessible' and 'remote' – adopted from the Scottish approach. However, when the 30-minute threshold is applied to Northern Ireland, only 7% of the population qualifies as remote. The report recommends that information at the 20-minute threshold also be included in definitions of rural, and 80% of the population is within this threshold. 'Within' or 'outside' the drive time zones was adopted as the standard language, in place of 'accessible' and 'remote' (NISRA, 2015).

There have been three phases to rural proofing in Northern Ireland – first in 2002, second in 2009, and third, the Rural Needs Act of 2017. Rural proofing has been around from the earliest stages of devolved government in Northern Ireland, first appearing as a commitment to 'routine consideration' of the 'rural dimension' in the 'making and implementation of policy' by all ministries (Northern Ireland Executive, 2001, p. 48). The DARD took on the responsibility of developing and publishing guidance (DARD, 2002). This mirrored the development of rural proofing in England. The Minister of Agriculture and Rural Development chaired the Rural Proofing Steering Group. The body was made up of a high-ranking representative from each department and was tasked with providing guidance and reviewing the effectiveness of rural proofing, at that time defined as (1) gathering information on the number and nature of policies subject to rural proofing, (2) specific provisions to meet rural service delivery, and (3) examples of best practice, all to contribute to publishing an annual report (Shortall and Sherry, 2018).

In the second phase, in 2009, rural proofing was enhanced in Northern Ireland. Despite recommendations on how to re-focus away from processes and towards outcomes by means of more independent monitoring and reporting (Northern Ireland Assembly Research and Library Services, 2009), the second iteration of rural proofing does the opposite: removing any reference to the cross-government steering group within the guidance document, and providing no indication of how the rural impact assessments are to be quality assured and signed off within each respective government department (DARD, 2011). The only potential recourse for challenge is the added expectation that 'Rural Impact Statements' will be made available as part of public consultations, as well as a suggestion to contact two named organizations to represent the needs of rural stakeholders (Shortall and Sherry, 2018).

A notable change in the scope and emotive characterization of key concepts and objectives is also evident. Initially, rural proofing is defined as examining policies "carefully and objectively to determine whether or not they have a different impact in rural areas" (DARD, 2002, p. 2) but becomes a 'proper assessment' to find the 'direct and indirect impact' on rural areas (DARD, 2011, p. 3). This expands the required *ex ante* analysis to include both expected and unexpected consequences – and gives the impression that a proper approach is actually subjective in that it keeps looking until some rural accommodation is identified. Evidence of an increasingly emotional and defensive characterization of rural proofing can also be identified by comparing the official guidance documents (Shortall and Sherry, 2018). Originally, the potential difficulty in achieving economies of scale in rural areas is fully acknowledged, and compensation with higher allowances per unit cost offered up as a logical and practical 'adaptation' option to address the delivery of government services in a rural context (DARD, 2002). However, within the enhanced guidance, the language and connotations associated with such an option change. The allowance of higher unit cost thresholds is labelled with the emotive term 'rural premium' and is presented as something to be avoided (DARD, 2011).

The third phase is the current one, signified by the introduction of the Rural Needs Act. Despite a commitment to undergo an independent evaluation of rural proofing (DARD, 2011), no evidence base was established on the effectiveness of previous and existing approaches before a ministerial decision was taken to introduce legislation in November 2015. No evidence base about access to rural services was used to inform the introduction of legislation about rural need. Around the same time as the subsequent Rural Needs Act (RNA) was granted royal assent (May 2016), a restructuring of government ministries came into force. Agricultural and rural development remained together, with some functions held by the previous Department of Environment moving to join them. The sponsorship of the RNA remained with the modified Department of Agriculture, Environment, and Rural Affairs (DAERA).

The RNA places a duty on public authorities to have 'due regard to rural needs' when 'developing, adopting, implementing or revising policies, strategies and plans, and designing and delivering public services'. The Act defines rural needs as 'the social and economic needs of persons in rural areas'. As passed, the responsible public authorities include government departments, local government, and a selection of non-departmental public bodies. Public authorities are obligated to compile information on how rural needs are addressed for inclusion in their own annual reports and for submission to DAERA.

Within the framework of the RNA, DAERA maintains an advisory role, continuing the historical approach by DARD before it of issuing guidance and training, but also taking on responsibility for compiling the information about how due regard to rural needs is being met across public bodies annually. While the Rural Affairs Department is tasked with laying the report before the assembly, and the rural affairs minister with making a statement, there is no clear indication of how, if at all, a formal 'watchdog' will be part of the monitoring and evaluation process. DAERA is also responsible for reviewing and potentially updating the schedule defining public authorities subject to the Act at least every three years, and for "securing cooperation and the exchange of information between public authorities" (Rural Needs Act Northern Ireland, 2016, p. 2).

Rural needs

The case study is based on research carried out in Northern Ireland during the year 2016 (Shortall and Sherry, 2018). The project was commissioned by DARD/DAERA to investigate and provide advice on the implementation of the RNA.

Three focus groups were formed, one for each of the three categories of public bodies subject to the RNA: Northern Ireland government departments (GDs), local government districts (LGDs), and non-departmental public bodies (NDPBs). The number of participants in each focus group ranged from 9 to 15. There were two GDs unable to attend the focus groups, and in these cases semi-structured interviews were carried out with nominated representatives. The discussion was prompted by asking what types of rural issues need to be considered in the context of their organization, how and if rural proofing had been undertaken previously, any lessons learned from their experiences, and what issues were being raised within their organization related to the introduction of the RNA.

An additional three semi-structured interviews were carried out with key informants to obtain perspectives from rural, environmental, and agricultural organizations. The conversations were initiated by asking participants how they felt their respective interests were addressed by rural proofing, how it could be improved, and how they understood the motivation and potential usefulness of the RNA.

Project information sheets describing the funding source, motivation, and objectives of the research, as well as contact information for the researchers involved, were provided to all participants. Written consent was also obtained. Focus groups and interviews were audio-recorded, transcribed, and analyzed by hand and with the software package MAXQDA (VERBI Software).

Although rural proofing started in Northern Ireland with governance at the ministerial level, each iteration has further delegated responsibility, at first within devolved government departments to civil servants and sponsored arms-length bodies, then further to public bodies more generally and local government. Concerns were expressed that the broadening of the scope and coverage of rural proofing by means of the RNA would ultimately be counter-productive because, in the words of one local government official, "if everybody has responsibility for it, it's diluted then" (LGD6). There is also concern that the rural ministry 'will hide behind the act' as their requirement to deliver for rural because 'everyone else is doing it' (rural organization).

There is a sense, particularly at the most local levels, that the devolution of rural issues is only at the surface because grass-roots and bottom-up innovations are stifled by the existing policy hierarchy, particularly in the case of rural development policy, whereby local strategies have to fit tightly within a national policy framework pre-emptively decided by DAERA, the same ministry that sponsors the RNA. For example, a rural development officer working within local government stated, "It's nearly the wrong way around...we go to implement our strategy but [it's] not eligible because it wasn't in the [national] program in the first place" (LGD2).

Some participants also felt that the elected members of the legislature were off-loading responsibility for improving rural service delivery on them, as opposed to empowering them to meet local challenges, particularly as additional responsibilities under the Act have no resource complement. For example, an arms-length body described it as, "Requiring us to deliver services that in theory they're legislating for. If they want them to be better for rural areas, they should try and legislate or set policy in that regard" (NPDB2). This perception of the Act is consistent with the analysis above that, while additional weight has been applied to the statutory duty, responsibility for delivering is actually diffusing outwards and downwards.

> I think the other challenge with all this stuff is that regardless of policy or anything else, unless you have a budget and there is scope to do things, it is very hard to…you can proof policies and all the rest of it but to make any kind of difference, you have to have some sort of resource of some description to make anything happen, particularly in the rural areas.
>
> (LGD3)

Considerable issues arise around broadening the scope of rural proofing not only in terms of responsibility, but also in terms of the fundamental concepts and intentions on which the RNA is constructed. Ambiguity regarding how to define 'rural' and the associated difficulties were raised within every focus group and interview. The failure to define rural clearly, leaving flexibility in the interpretation by public bodies in the course of compliance, is seen as a weakness rather than strength of the approach. This is because there is no common understanding of "who people are and what business we're serving and what needs we're targeting: (GD3), meaning that there are grounds for "ministers stepping in and stopping" plans in place that "professionals had sorted" (NDPB5), or for legislators to challenge in committee what civil and public servants consider as a "fairly scientific process to get at what was rural" (GD5) if it didn't align with how they viewed their own constituency. Instead of affording practical flexibility to redefine rural to suit specific contexts and cater to the most appropriate evidence given that context, the open treatment of rural tends to create further space for subjective and emotional factors to muscle in.

> It's not about facts. It's about narratives. It's about framing. It's about values. It's about the heart. It's about perception. It's not about the facts at all.
>
> (Environmental organization)

The amorphous treatment of 'rural' is only one dimension of a fundamental hurdle to applying such an approach to gauge the effectiveness of rural service delivery – namely, that the cornerstone of the RNA, to ensure 'due regard' to 'rural needs', relies on vague and emotive language. From the very title, it associates 'rural' with 'need' then fails to specify what 'rural needs' are and how they differ from the 'social and economic needs' of people generally: the resulting connotation is that of 'rural neediness'.

> It says here…the reasonable provision to meet needs, and, you know, when does a need become a want? …what are rural issues, and what's the evidence to back up that rural issue? Where do you stop?
>
> (GD3)

From a user point of view. . .some of the comments we're getting back. . .that it's my human right to have a social security office along the road where I live. . .There seems to be, what came first, rural needs or a rural impact, or a Rural Needs Act, or rural impact assessment? . . .It's almost creating its own demand.

(GD1)

The result is a dissonance between the duty as described (addressing rural needs) and how public service practitioners understand the problem: how to deliver against common needs in a rural context. While several public bodies referenced, or were comfortable with, the concept of due regard to 'space' (spatial impact assessment, spatial analysis, and geographic-information-system-derived evidence), the duty to 'rural' proved more elusive and difficult to translate into practical accommodation.

So. . .we do recognize that there are inequalities in rural areas and particular issues that are unique in rural areas. . .such as isolation, social isolation, isolation from services. . .an aging population. . .they're not specifically rural, I mean you get poor people in urban areas as well. . .and you get social isolation, but. . .the interventions probably need to be different because there may be spatial issues. . .it's at the delivery level that it's, probably, you know, more relevant.

(GDInt2)

In addition to concerns regarding the reinforcing or even fostering of perceptions of rural entitlement, and confusion regarding how to translate the RNA into service delivery models, another poignant observation emerged: that the language used may even cause offense to people who identify themselves with rural.

Well it almost like then. . .you're gaming the system or you're priming it already to be. . .a welfare recipient model. . .you've already built in, rural by definition is you're a victim, you're a need, you're vulnerable. . .I think it's disparaging, it's inaccurate. . .to make this a public badge, rural equals neediness. . .like describing women as always victims.

(Environmental organization)

This is a particularly interesting comment about rural equating neediness, while in Northern Ireland the population in rural areas consistently increases while urban population declines (Russell, 2015). In other words, people are choosing to live in rural areas.

Rural proofing as a method to ensure service delivery

This chapter considers the potential of rural mainstreaming models as a method to evaluate the success of, and improve where possible, rural service delivery.

The case study examined is the introduction of novel legislation that transposes the English model of 'rural proofing' to a duty of 'due regard' to 'rural needs'. The evidence suggests that the scope and reach of rural proofing in Northern Ireland have progressively expanded over time; however, this has led to a wide and thin allocation of both responsibility and expectations for related outcomes. The fuzziness around outcomes is further exacerbated by the fact that there is no coherent evidence base identifying the problems facing rural service delivery. The establishment of a shared understanding across public bodies of rural priorities and challenges is further complicated by the treatment of foundational terms and concepts within the legislation. The choice of language favors the intangible and the difficult-to-measure. Therefore, a bias towards subjective determinations and challenges of what constitutes rural, rural needs, and due regard is observed or expected by those impacted by the new duty. This potentially foments underlying dissent amongst stakeholder groups, public bodies, and the individuals within them on what constitutes 'rurality' rather than fostering cooperative and innovative solutions across public services and administrative jurisdictions.

Note

1   The Rural Needs Act (Northern Ireland) 2016 received royal assent on 9 May 2016. It requires government departments, local councils, and other public bodies to ensure the 'social and economic' needs of rural areas be given 'due regard' in all strategies, policies, and plans including the delivery of services.

References

Atterton, J. 2008. *Rural proofing in England: A formal commitment in need of review.* Discussion Paper Series, No. 20. University of Newcastle Upon Tyne: Centre for Rural Economy.

Bock, B.B. 2004. "Fitting in and multi-tasking: Dutch farm women's strategies in rural entrepreneurship", *Sociologia Ruralis* 44(3): 245–260.

Bryden, J. 2009. "From an agricultural to a rural policy in Europe: Changing agriculture, farm households, policies and ideas". In P.J. Stewart and A.J. Strathern (eds.), *Landscape, heritage, and conservation: Farming issues in the European Union* (pp. 141–168). Durham, NC: Carolina Academic Press.

Bryden, J., Efstratoglou, S., Ferenczi, T., Knickel, K., Johnson, T., Refsgaard, K., and Thomson, K. 2010. *Exploring inter-relationships between rural policies, farming, environment, demographics, regional economies and quality of life using system dynamics.* Oxfordshire: Routledge.

Copus, A., Hall, C., Barnes, A., Dalton, H., Cook, P., Weingarten, P., Baum, S., Stange, H., Lindner, C., Hill, A., Eiden, G., McQaid, R., Grief, M., and Johansson, M. 2006. *Study on employment in rural areas (SERA).* Brussels: DG Agriculture, European Commission.

DAERA. 2017. *A guide to the Rural Needs Act (NI) 2016 for Public Authorities.* Belfast, NI: Department of Agriculture, Environment, and Rural Affairs.

DARD. 2002. *A guide to rural proofing: Considering the needs of rural areas and communities.* Belfast, NI: Department of Agricultural and Rural Development.

DARD. 2011. *Thinking rural: The essential guide to rural proofing*. Belfast, NI: Department of Agriculture and Rural Development.

DARD. 2015. *Public consultation on policy proposals for a rural proofing bill*. Belfast, NI: Department of Agriculture and Rural Development. Available online at www.dardni.gov.uk/index/consultations/rural-proofing-bill-consultation-2015.htm Accessed November 12, 2015.

Department for Environment, Food and Rural Affairs. 2015. *Independent rural proofing implementation review*. Lord Cameron of Dillington: Department for Environment, Food, and Rural Affairs. Available online at www.gov.uk/government/uploads/system/uploads/attachment_data/file/400695/rural-proofing-imp-review-2015.pdf.

DSD. 2016. *Family resources survey urban rural report Northern Ireland 2013–2014*. Available online at http://dera.ioe.ac.uk/26456/1/frs-urban-rural-report-ni-1314-full-copy.pdf.

Jack, C.G., Anderson, D., and Patten, N. 2012. *Rural households' experience of accessing public services in Northern Ireland*. Belfast, NI: Agri-Food and Biosciences Institute.

Knox, C. 2008. "Policy making in Northern Ireland: Ignoring the evidence", *Policy & Politics* 36(3): 343–359.

NISRA. 2005. *Report of the inter-departmental urban-rural definition working group*. Belfast, NI: Northern Ireland Statistics and Research Agency.

NISRA. 2015. *Review of the statistical classification and delineation of settlements*. Belfast, NI: Northern Ireland Statistics and Research Agency.

NISRA. 2017. *Northern Ireland summary statistics*. Belfast, NI: Northern Ireland Statistics and Research Agency.

Northern Ireland Assembly. 2016. *Rural Needs Act Northern Ireland*. C. 19. Norwich, UK: The Stationary Office.

Northern Ireland Assembly Research and Library Services. 2009. *Rural proofing and the rural champion*. Available online at www.niassembly.gov.uk/globalassets/documents/raise/publications/2009/agriculture/3109.pdf.

Northern Ireland Executive. 2001. *Draft programme for government*. Available online at http://cain.ulst.ac.uk/issues/politics/programme/pfg2001/pfg2001d.pdf.

OECD. 2011. *OECD rural policy reviews*. England, UK: OECD Publishing.

Russell, R. 2015. *Key statistics for settlements, Census 2011*. Research paper 99/15. Northern Ireland Assembly: Northern Ireland Research and Information Service.

Sherry, E. and Shortall, S. Forthcoming. "Methodological fallacies and perceptions of rural disparity: Real versus abstract needs". Submitted for publication.

Sherry, E. and Wu, Z. 2018. "Can rural development policy improve employment income inequality? Using structural patterns to understand the trade-offs between spatial and non-spatial groups". Submitted for publication.

Shortall, S. 1996. "What are the new approaches to rural development?", *Economic and Social Review* 27(3): 286–305.

Shortall, S. 2012. "The role of subjectivity and knowledge power struggles in the formation of public policy", *Sociology* 47(6): 1088–1103.

Shortall, S. 2013. "Sociology, knowledge and evidence in rural policy making", *Sociologia Ruralis* 53(3): 265–271.

Shortall, S. and Alston, M. 2016. "To rural proof or not to rural proof: A comparative analysis", *Politics and Policy* 44: 35–55.

Shortall, S. and Sherry, E. 2018. *Rural proofing in Northern Ireland: An overview and recommendations on guidance, implementation and governance*. Belfast: Agri-Food and Biosciences Institute. ISBN 978-1-908471-08-6. Available online at www.afbini.gov.uk/publications/rural-proofing-northern-ireland-overview-and-recommendations.

Shortall, S. and Shucksmith, M. 2001. "Rural development in practice: Issues arising in Scotland and Northern Ireland", *Community Development Journal* 36(2): 122–134.

Shucksmith, M., Thomson, K., and Roberts, D. eds. 2005. *The CAP and the regions: The territorial impact of the Common Agricultural Policy.* London: CAB International.

Sociologia Ruralis. 2000. "Special issue: The EU LEADER programme". 40(2).

3 Service Tasmania

Australia's first whole-of-government initiative

Greg Blackburn

Introduction

Government service delivery in Tasmania, Australia, has traditionally been difficult for those living in rural and remote areas due to geographical distances. Access to government services for Tasmanians has previously required visits to various lead agencies at different locations, as government service delivery was fragmented and driven by inefficient bureaucratic processes. In this context, this chapter describes a major public-sector reform project promoting a whole-of-government, customer-focussed approach to service delivery.

Two decades ago, the relationship between the citizens of Tasmania and their government fundamentally changed, undergoing a customer-orientated shift. Public administration service delivery in Tasmania undertook a radical new direction with an innovative and creative service-delivery model which was guided by 21st-century principles and was particularly beneficial to rural areas and small towns. Political and external forces were the main drivers. The response was the birth of Service Tasmania, a single integrated one-stop approach to providing seamless, cross-agency government service delivery.

Today, Service Tasmania offers almost 600 government services to the community through physical, online, and telephone channels and undertakes more than 1,500,000 service transactions annually. This chapter will reflect on the lessons learned through Australia's first whole-of-government initiative. The benefits which rural and urban communities obtained through this customer-orientated shift in government service delivery, the catalyst for change, and implementation challenges are illustrated. Australia's first whole-of-government initiative can, in summary, be applauded as a success and serves as a good example of how state-of-the-art technology can facilitate innovative citizen-centred government and spur economic growth.

The Service Tasmania context

Government service delivery has transformed significantly over recent decades. Many have witnessed a wave of public-sector reform projects internationally which modernize, and radically restructure, the public sector toward efficiency and citizen-centric government service delivery. Such projects are changing the landscape of government service delivery, largely benefiting rural and remote locations. The Australian Bureau of Statistics' definition of 'rural' is a population cluster of less than 1,000 people (Sher and Sher, 1994). People living in urban areas also receive many benefits, though the impact on them is not as great as their rural counterparts, as they typically already have easy access to such services. Governments and councils are moving beyond a narrow concentration on 'core' services, 'roads, rates, and rubbish,' commonly said to be the three top concerns of local government, towards an emphasis on broader objectives such as promoting the economic, social, environmental, and cultural well-being of local communities (Binning and Young, 1999; Inayatullah, 2011; Johanson *et al.*, 2014).

Traditionally, government service provision was bureaucratic and poorly integrated, if at all, requiring visits to various agencies to access different government services. Not only was this inconvenient, it caused delays, and the delivery overheads themselves were expensive. Residents and businesses in remote locations were acutely aware of these time and financial discomforts as simply being a fact of life ever since government agencies first opened their doors. However, advances in digital technologies offer many advantages leading to governments transforming their service-delivery models, consequently reshaping government service-delivery contexts. Technology continues to impact how people work, play, gain information, and participate in communities, and it is changing the way we interact with government agencies and their service delivery.

With the introduction of technology, government service delivery in Tasmania has undergone a radical change, altering the nature of the relationship between the government and the community. Service Tasmania offers new possibilities because of its whole-of-government approach to service delivery. This facilitates cross-agency, customer-targeted service delivery within the overall government framework. The Service Tasmania project was the first whole-of-government cross-agency initiative in Australia (Blackburn, 2016). This government service-bundling initiative is widely considered to be a successful, well-organized, well-managed, and well-implemented project (Blackburn, 2014). And it has demonstrated two decades of high-quality service, making it easier for Tasmanians to access services, especially those in rural and regional areas (Croger Associates *et al.*, 2003). Therefore, this example of public-sector reform provides substantive insights for public service administrators, policy makers, and practitioners who are engaged, or interested, in service-delivery reform projects.

The Tasmanian setting

Tasmania, Australia (see Figure 3.1), the world's 26th-largest island, started life as Britain's prime penal colony housing around 75,000 convicts. It is known for its vast, rugged wilderness areas, mountain ranges, and 4,000 lakes in the Central Plateau (akin to that found in northern Canada and Finland). Being the smallest of Australia's six states, Tasmania has a land area of 68,401 square kilometres (26,410 square miles) and is 315 kilometres (195 miles) across and 286 kilometres

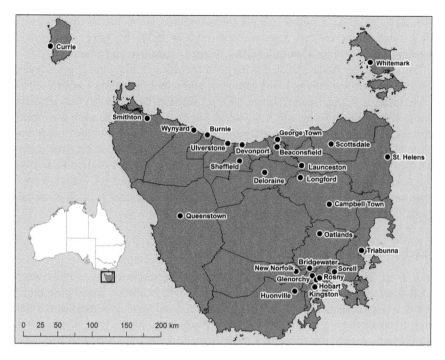

Figure 3.1 Map of Tasmania – Service Tasmania's 27 shop locations (Service Tasmania, 2018)

Map credit: Aita Bezzola

(177 miles) long (Brand Tasmania Council Inc., 2014). Comparatively, it's about one and a half times the size of Switzerland.

As of June 2016, Tasmania's population is 519,100 (Australian Demographic Statistics, 2016), divided approximately equally between the North and the South. About 40% of Tasmanians live in and around Hobart, the capital city. There are 90 small towns in Tasmania with populations under 5,000 people, 66 of which have populations under 1,000 inhabitants (Bourne *et al.*, 2017). Some rural areas lack basic local services and infrastructure, such as accessible education, government services, healthcare, adequate and affordable drinking water, employment, transport, public telephones, and broadband internet connection (Tasmanian Council of Social Service, 2009). Tasmania has the highest proportion (31%) of households dependent on government pensions and allowances in all states and territories (Adams, 2009; Department of Premier and Cabinet, 2009).

Tasmania's geography is distinctive because the proportion of mountainous country to total area is particularly high (About Australia Pty. Ltd., 2017). When travelling through remote or wilderness areas, the roads become narrow and winding, and are often gravel. Distances shown on maps can be deceiving as many roads wind up and down the sides of mountains and hills, protracting travelling times. In addition, wildlife adds further complications and dangers to travel. Consequently, while rural towns are not separated from major cities by great geographical distances, they are somewhat isolated due to protracted travelling times.

The geography has traditionally led to difficulties in integrating government services and service infrastructure into rural and remote locations (Adams, 2009; Department of Premier and Cabinet, 2009). Likewise, the insularity has rendered much of its political, economic, and social life distinctive. To access government services that were in disparate establishments, Tasmanians have previously had to travel to major centres, often hours away. The absence of train services means that Tasmanian citizens rely on bus, coach, or private vehicles to get to and from cities and regional centres. Social isolation is also a concern for many communities in rural and regional areas, as this can impact negatively on physical and mental health and well-being (Adams, 2009).

Catalyst for change

Prior to the Service Tasmania initiative, the tyranny of distance used to afflict many small businesses and citizens from rural townships who needed to access government services. Geographical isolation is felt in many rural and regional Australian communities (Apperley *et al.*, 2011; Blainey, 1982; Ward *et al.*, 2007). Thus, simple things like paying a speeding infringement meant travelling for hours to the nearest major town. This caused not only disruption and inconvenience for businesses missing employees who had to travel far to access government services, but also caused people the embarrassment of having to tell their employer the nature of their absence. This is one simple illustration. However, major economic and social effects due to the difficulty in providing integrative services and service infrastructure were felt in many Tasmanian communities. In responding to calls to improve service delivery and economic development, the then Tasmanian Premier, Premier Rundle, stated in 1997 that sub-national economies were struggling to provide the wealth and employment needed by their people. Premier Rundle identified four primary economic concerns of Tasmania: industries were reducing their workforces as they struggled to remain competitive, national government services had been cut back or withdrawn to national or state capitals, banking and private-sector services were withdrawn to larger population centres, and young people were leaving to find employment interstate.

The minority Rundle government executed a reform agenda aiming to catapult Tasmania from its traditional economic and social malaise to a better place (Rundle, 2014). Along with economic concerns, a need for making it easier for Tasmanians, especially in rural and regional areas, to do their everyday business with government was recognized.

Service Tasmania concept

The Service Tasmania concept is based on the New Brunswick model developed in Canada (Blackburn, 2016; Rundle, 1997). Service New Brunswick (SNB) is the provincial government's chief provider of front-line services to the public, first established in 1992 (Kohlborn *et al.*, 2010; Service New Brunswick, 2012). SNB is a corporation owned by the Province of New Brunswick with a mission to improve the delivery of government services to the public. Acting as a gateway involving public-private partnerships, SNB has a network of 49 one-stop delivery shops throughout the province. The services that SNB provides include a wide range of geographic and land information products and services, and access to more than 120 government services provided on behalf of various government departments and agencies. SNB is instrumental in turning the province into one of the fastest-growing regions in Canada (Vincent and Prychodk, 2002).

The SNB model was found to be suitable for Tasmania as the Canadian environment has a similar economic climate and socioeconomic environment, and the youth were also leaving the region to seek employment. Parallels were observed between Tasmania and New Brunswick in terms of size and population, rural composition, economic development, and telecommunications infrastructure. SNB presented convincing evidence that the use of advanced telecommunications and information technologies could transform a regional economy.

Whilst the New Brunswick model was the foundation for the vision, it was adjusted to meet regional demands, making the Service Tasmania model uniquely Tasmanian. The Service Tasmania model resembles Service Canada (Fraser, 2009; Reddicka and Turner, 2012), Service Ontario (Borins, 2003), and New York City 311 (Sharma *et al.*, 2014) in that each organization eliminates existing agency silos. Instead, they offer one-stop multichannel services spanning three customer-access channels (telephone, digital, and face-to-face), and employees have multiple functions across the organization. The result is a whole-of-government approach to service delivery, facilitating seamless cross-agency customer-targeted service delivery within the overall government framework (local, state, and federal) (Barrett and McFarlane, 2002). The key objective of Service Tasmania is to enable the Tasmanian community to locate and interact with government services and information without the need to know which level of government, or which department or agency, provides the information or service they seek (Blackburn, 2016; Sokvitne, 2002). The model was proposed as the key to generating a strong Tasmanian economy (Rundle, 1997). Service Tasmania represents a major reform in public-service delivery, challenging the traditional model of public administration through the application of private-sector-style management practices and digital technologies.

As a government services portal for Tasmanians, including commonwealth, state, and local government, plus private-sector business services, Service Tasmania breaks through existing agencies' silos to offer one-stop multichannel services. The one-stop-shop concept, described by Reid and Wettenhall (2015) and Shah (2001) as improving communications between service providers and the populations they serve by eliminating bureaucratic stumbling blocks, makes access to government easier, faster, and more convenient, especially for rural communities, as it overcomes the geographic centralization concerns mentioned above. Operating on the 'one entry point to government' model, Service Tasmania uses the principle that people do not necessarily need to know what level of government they are dealing with.

By bringing common government services together under one organization and embracing private-sector business practices, Service Tasmania pursues more effective and efficient ways of delivering traditional services. This, mixed with a stronger focus on client needs, has led to economies of scale and customer-centric service delivery, a major policy and legislative change which also represents changes in managerial philosophy. Further, in keeping with private-sector business management principles, employees have multiple functions across the organization, representing a major ethos change in governmental operations. Failing rural economies had witnessed several businesses file for bankruptcy, resulting in people leaving these areas in search of employment elsewhere. Service Tasmania has provided wealth-generating abilities to rural communities, resulting in employment opportunities and creating population stability that provided the foundation for community well-being.

Today, Service Tasmania's 27 one-stop shops offer 599 services, plus an array of telephone and internet services. Service Tasmania carried out more than 1,600,000 transactions over the counter (906,289 of these were financial transactions with AU $181,577,891 revenue being collected), answered almost 230,000 calls in the contact centre, and facilitated more than 300,000 electronic payments over the phone and internet in 2013–2014 (Burton, 2014). Rural and small towns have benefited from the streamlining of service delivery and from easier access to more services provided to citizens and local businesses, regardless of their location.

Structure, infrastructure, staffing, and shops

Service Tasmania represents a paradigm shift in government service delivery; as such, the organizational impacts of this public-sector reform are widespread, both on the community and the structures of service delivery, as the Tasmanian government reinvented itself. The following is an overview of the generic Service Tasmania model.

Service Tasmania structure

Service Tasmania is administered by the Service Tasmania Board, which is chaired by the Deputy Secretary, Department of Premier and Cabinet. The Board consists of the Departments of Treasury and Finance, Environment and Land Management, Transport, Primary Industries and Fisheries, client agency representation, and a position appointed by the premier from outside government. The business unit reports directly to the Board, a unit from within a lead agency where Service Tasmania's expenditure and income is accounted for. There is a designated business unit manager who is responsible for managing the delivery of Service Tasmania services including technology, human resources, finances, and maintenance of infrastructure and assets.

Two agency types exist: 1) lead agencies responsible for the delivery of services and 2) client agencies that have their services delivered through Service

Tasmania by a lead agency. The business unit's duties include acting as a 'backroom' or a point of contact (support unit) for the service shops, to assist with technical, operational, and banking services, and to accept and process daily information to the office on all banking and business transactions. The day-to-day management of the shops is supported through the adoption of a regional approach, which has been divided into three regions: 1) northern, 2) southern, and 3) northwest. A regional manager is responsible for eight shops in his or her region. The advantages of the regional concept are that it provides management on a local basis nearby; closer working arrangements between the rural and urban shops, which fosters a mentoring arrangement among staff; and a local approach to staffing requirements in terms of on-the-job training, quality assurance, and relief arrangements. Customer service officers (CSO) are responsible for the front-line delivery of government services, including accepting government receipts, performing transactions, and providing customers advice and information on government departments.

Infrastructure

Service Tasmania utilizes information and communications technology (ICT) for transactions between the government and its customers, and communication amongst its shop network. In implementing the one-stop shops, Service Tasmania was initially restricted to the existing level of technology employed by the participating government agencies. IT services 'bundled' the various interfaces from the many lead agencies into one platform, which had a number of web-based applications to assist.

Due to the limitations and constraints placed on the implementation of the Service Tasmania generic model, pre-existing client agency technologies were utilized. These limitations included time and finances. Existing technologies came from government departments and were not specific to Service Tasmania's operations. Subsequent technology issues negatively affected rural service delivery, as staff had to use different systems, often leading to confusion, frustration, breakdowns, and errors. This negatively impacted on service delivery, resulting in poor customer relations.

The IT environment comprises three infrastructures: one each for the business unit, the rural shops, and the urban shops. The Telstra public telecommunications network infrastructure is used as the framework for delivering services and supplying information between shops. All CSOs have access to networked computers and a uniform package of office equipment. The systems employed belong to existing government departments and are consistent across all shops.

Staffing arrangements

The recruitment and training of staff were completed per the requirements of the lead agencies. Staff were recruited from existing government staff and from the public. Staffing for rural and urban shops was undertaken via different recruitment drives. As some lead agencies were opening new rural-area shops, they could advertise and run a detailed induction program. During these programs, staff were told Service Tasmania's philosophy, what was going to happen (an expectation to complete large volumes of transactions, provide information to the public, be customer orientated, etc.), and what services they would offer. In some instances, due to the expedited project roll-out, it was not known exactly who the staff would be until only a week prior to a shop's opening. Prior to shop openings, information sessions were held, where staff were told what they would be doing and advised on the status of the shop development, such as renovation progress. Depending on whether a rural roll-out or an urban roll-out was planned, training was either well planned and executed, or rushed and not thought through. Often urban rollouts suffered from hastened implementations.

Transfers from existing agencies helped fill vacancies quickly but led to some subtle differences in conditions of employment; new agreements rectified this in time. The staffing arrangements made provisions for two or more CSOs to work simultaneously in all shops. Staff facilities consisted principally of a modern customer service desk staffed by two to three CSOs in a rural shop or 20 in a larger shop. All staff have access to the various client agency systems.

The one-stop shop

A generic shop model was formulated by Service Tasmania and has been used for all state-wide locations. Importantly, this model allows for localization 'amendments' on a 'needs be' basis. Such amendments are applied, for instance, when co-locating with other government and non-government organizations, for example police, charity, and banking organizations. Having a generic model in place where similarities can be found at all locations means that citizens can go into any shop around the state and expect to find a recognizable environment and gain access to the same range of services to complete their business. A full range of services exists in each shop, though some offer additional services, for example access to banking facilities and human services or, more commonly, Centrelink, which is a one-stop, multi-purpose delivery agency (Halligan and Wills, 2008; Norman, 2010; Vardon, 2002) providing services to people in severe financial hardship and often in crisis (Hall *et al.*, 2016). Service Tasmania acts as an agency for Centrelink, providing commonwealth and state government services (Barrett and McFarlane, 2002; Szirom *et al.*, 2001).

Where co-location arrangements have been made with other government and non-government organizations, poor and good relationships have been experienced. Corradini *et al.* (2018) argue that co-location is difficult due to shared resources and culture differences. Where poor co-location relationships have existed, the resources needed to facilitate both parties have been lacking. This has led to increased tensions and stressed dynamics. Similar issues are illustrated by Grace and Coventry (2010), who argue that organizations contemplating co-location need to build on pre-existing trust and goodwill when planning for potential co-location impact, consulting with stakeholders, managing change, and acknowledging the different cultures. In the Service Tasmania experience, where adequate resources are provided, good relationships between co-location colleagues exist which often develop into friendships, phenomena illustrated by Frost (2005). In these shops, the CSOs and their co-location peers often share each other's tasks.

Implementation and challenges

Whilst the Service Tasmania project ultimately has been successful, it was subject to several impediments that threatened to derail the project in the early stages. Human resources are critical to the success of any endeavour; therefore, it is essential that they perform the function in a manner affording benefit or advantage to the organization. Service Tasmania did encounter some early difficulties with some staff members due to recruitment limitations. Staffing issues included inadequate training, resistance to change, isolation, and information overload. Service Tasmania managers actively addressed each, and a dominating commitment to the organization developed.

Time and financial limitations placed on the project's implementation meant pre-existing computer technologies were utilized, staff recruitment was hurried, and premises were occupied quickly. This led to several negative issues as staff training was initially rushed, leaving many staff feeling inadequately trained. Where co-location stresses occurred, they led to petty rivalries and culture clashes. Staff frustration and confusion resulted from the various systems, functions, and keystrokes required to operate inconsistent government software programs.

Staff recruitment from various agencies led to co-workers with the same job titles, functions, and roles operating under different employment conditions. This affected salaries, maternity leave, overtime, etc. Unfortunately, this sometimes led to industrial complications and union involvement. Interestingly, this issue occurred mostly in rural areas, as this was where the majority of new shops were opened. Over time this issue was resolved by implementing uniform employment conditions and contracts. Bringing people together from various organizations originally meant a blend of cultures, sometimes not an ideal situation. Service Tasmania managers made deliberate attempts to embed a new corporate culture organization-wide and make it 'stick' so that employees did not slip back into old habits, norms, and values. Actions recognized by Barratt-Pugh *et al.* (2013) as important, otherwise embedded traditions and behaviours could re-emerge as the dominant culture.

Training has been one of the largest issues that evolved out of the project. The initial quick implementation time and budget limitations resulted in reduced training opportunities and numerous staffing concerns. Many CSOs expressed discontent about their lack of training. Rural staff were given two weeks of intensive training. Urban staff were not exposed to this extensive training due to the quick changeover of operations.

Resistance to change is often encountered during organizational reform projects (Blackburn, 2014; Bovey and Hede, 2001; Day *et al.*, 2017; Georgalis *et al.*, 2015), and there were many examples occurring during the Service Tasmania roll-out. Integrating services from several agencies and organizations requires communication and support across each affected organization. Otherwise, resistance in the form of bottlenecks in service delivery by one or more agencies can occur. The root cause of resistance occurrences was identified and addressed by management. Supporting staff through training and development, plus informing and supporting them on changes to their daily work, helped address barriers and was a critical element in successfully delivering the change strategy.

Communication technologies played a large part in addressing staff feelings of isolation. Managers can now immediately broadcast information to all staff, groups, or individuals. This particularly helps rural staff, allowing them to seek clarification from individual management or ask for help from groups or colleagues. Technology has reduced the distance barriers previously hindering communication. Effectively communicating the vision, the mission, and the objectives of the change effort has helped CSOs understand how they are personally affected, subsequently reducing their resistance.

A detailed cost-benefit analysis has not been undertaken, or has not been found so far. Apart from benefiting rural areas economically, there are also indications of economic benefit to the organization (reduced service-delivery overheads, etc.); however, there appear to be few informative and strict evaluations. Notwithstanding, figures from the Service Tasmania 2014 Annual Report (Burton, 2014) present a total receipt carry forward amount of more than AU$880,000, suggesting that agency, partnership, and other fees are more than compensating all budget expenditure. The move to a business-centric business administration model is designed to modernize and radically restructure the public sector towards efficiency and profitability. So far, the initiative appears promising.

Transferable lessons from Service Tasmania

Service Tasmania provides insights into a delivery model that has been successfully applied in practice. Identified lessons from the Service Tasmania model and its execution are of benefit to other government reform projects, are widely transferable, and could be used as a starting point for both policy and practice for other governments. Although the Service Tasmania model is uniquely Tasmanian (McCann, 2001), it involves the application of private-sector business practices and ICT to deliver a wide range of government services to a broad geographic area. Unsurprisingly, technology issues arose throughout in the beginning. Patched-together back-end systems used quickly modified front-end interfaces. System inconsistencies caused numerous errors, mistakes, and frustrations. Bundling existing government technologies led to staff dissatisfaction, poor service delivery, and technical difficulties (Cresswell *et al.*, 2006; Nam and Pardo, 2013). Therefore, developing uniform Service Tasmania-specific systems has improved usability and addressed these issues.

A significant uniqueness to the Service Tasmania model is that service delivery is not solely provided by government agencies. Service Tasmania acts as an interface to the provision of services from local councils, various state and commonwealth (federal) government agencies, and some private-sector businesses (banking sector, not-for-profit community services, etc.). The one-stop delivery model provides a 'value for money' service and gains high economies of scale, as financial systems, website maintenance, and development activities are now centralized. In addition, the model allows for 'localized customizations' and achieves a high degree of design consistency having all services delivered under the same brand. Importantly, such government entry points must originate from the citizen's perspective rather than that of government agencies to be relevant (Dugdale *et al.*, 2018).

However, co-location stress was demonstrated at various times in the project. Stress concerns in co-locations arose where facilities were inappropriate or the requirements of all parties were unmet. Extensive shop renovations assisted co-location relationships by addressing the needs of both parties. Where successful, co-existing staff often learned the jobs of their counterparts to provide backup when necessary.

Adding time and financial resources overcame the training issues that Service Tasmania experienced. Organization-wide training concerns were addressed by providing extra training sessions and by implementing a staff-training centre, buddy systems, and tailored courses. This remediated the deficiencies encountered in the implementation – lessons others can learn from.

Staff recruitment issues due to subtle differences in conditions of employment were overcome with new agreements. Policy makers and practitioners managing public-sector service-delivery reform projects can learn from this experience by preparing such a universal employment agreement from the outset.

Expecting resistance to change is sensible. We know that the reasons for resistance to technological change are a loss of status or power, uncertainty, self-interests, lack of understanding and trust, different assessments and goals, changes in job content, economic insecurity, altered interpersonal relationships, and altered decision-making approaches (Boyd, 2011; Carter, 2008; Lewis, 2011; Oreg, 2006). The human element is a critical factor in any endeavour, and Service Tasmania CSOs have developed a powerful commitment to the organization and their customers. On several occasions, CSOs have gone beyond the bounds of their position description to serve their customers, demonstrating their commitment to the organization and a willingness to take the initiative. The rethinking of service-delivery processes, behaviour, and ethos is necessary with technology implementation and requires well-understood change-management interventions (communicate, celebrate wins, etc.) to overcome major administrative challenges or threats to achieving successful outcomes.

Whilst a study investigating the community benefits has not been undertaken in a rigorous fashion, anecdotal evidence suggests that significant benefits for rural and urban communities have been obtained. Utilizing ICT positively affects service provision capability to rural and remote areas. Technology aids in simplifying processes and improves front-end government service delivery. Perhaps the most significant indicator is the CSO perception that the community has greatly benefited from the Service Tasmania project, adding considerably to rural Tasmanian communities' well-being.

Service Tasmania has addressed the issue of service delivery around the state by employing digital technologies and implementing the one-stop shops using private-sector management principles, resulting in flow-on social and economic benefits. Previously, rural communities were forced to travel to a major centre to access government services, an arrangement that was inconvenient and costly to many. People can now access any service in any service shop and they can receive information on a wide range of government services online. Service Tasmania has effectively taken business out of the city and placed it in the country, putting services back into rural areas. Rural communities that had experienced a continual loss of services have benefited greatly from access to government services needed to generate a strong economy. The benefits for urban Tasmanian areas, though, have not been expressed as soundly as those in rural communities, reflecting the acceptance from urban people accustomed to having access to a multitude of services.

For public-sector managers involved in service reform implementation, a key factor in driving the achievement of such projects is the vision, which needs to be articulated to, and adopted by, all staff. Employee training and the demonstration of a clear vision are significant factors affecting employee motivation (Kim, 2014). In addition, accountability and governance (documented practices, job descriptions, standard operating procedures, universal business process models, etc.) are important in providing a consistent framework to support any post-implementation review activities (in assessing the effectiveness of the project in achieving stated benefits, etc.). Both positive and negative lessons, i.e., evidence of what did and did not work well, are of equal benefit and importance to this case study.

Summary

Advances in digital technologies offer us many advantages and have led to governments transforming their service-delivery models and improving the economic viability of many remote communities. Under pressure from external threats, the Rundle government utilized new technologies to establish Service Tasmania. The Service Tasmania concept has made significant contributions positively affecting Tasmanian citizens. Both rural and urban communities have benefited, with more visible positive effects being demonstrated in rural communities that had suffered from the continual removal of services and economic decline. These benefits are achieved by changes in managerial philosophy, incorporating private-sector-style business administration values, and providing extra service-delivery opportunities in rural communities, resulting in economic recovery and development. By bringing government services to rural communities, additional choices are afforded them. The public can conveniently access a range of services (obtain government information; complete their government business; purchase government products; pay taxes, fines, and government fees) in a single location near them.

Although Service Tasmania can be declared an overall success, it did experience many challenges that could have been fatal to the project's implementation. Project limitations of time and finances meant that staff recruitment was hurried, premises were occupied quickly, and pre-existing computer technologies were utilized. This led to several negative issues, as co-location sometimes led to petty rivalries and culture clashes. Staff training was initially rushed, leaving many staff feeling inadequately trained, and various inconsistent government software programs caused confusion and frustration. Implementing the model has demonstrated, for example, that if all parties in co-location relationships are not accommodated, difficulties do occur. When the needs of all parties are addressed, good relationships between staff can develop. The transferable lesson is, with time and appropriate finances, management interventions can resolve these described impediments.

Understanding the methods applied by Service Tasmania managers in implementing the service shops is important, as public-sector administrators globally can learn from the factors that led to the success of this change effort. Managing organizational change presents a minefield of politics, principles, and technical and people issues that often collectively account for much of the project. The Service Tasmania project has been successful because management challenged conventional ways of working within bureaucracies, whilst at the same time rigorously following guidelines that kept the project on track, within budget, and on time.

For public service administrators and others interested in the field, this chapter has provided substantive insights, as no major reform project is without its 'hiccups' and the Service Tasmania project was no different. However, challenges were met by determined management who not only managed the finances and tight project roll-out scheduling, but also considered the 'soft' elements. All change projects, ultimately, involve people – and it is people who are the key to an organization's success.

References

Australian Demographic Statistics. 2016. Available online at www.abs.gov.au/ausstats/abs@.nsf/0/BCDDE4F49C8A3D1ECA257B8F00126F77?Opendocument. Accessed March 9, 2017.

About Australia Pty. Ltd. 2017. Tasmania facts. Available online at www.about-australia.com/tasmania-facts. Accessed March 11, 2017.

Adams, D. 2009. *A social inclusion strategy for Tasmania*. September. Available online at www.dpac.tas.gov.au/__data/assets/pdf_file/0005/109616/Social_Inclusion_Strategy_R Repor.pdf. Accessed March 3, 2018.

Apperley, T., Nansen, B., Arnold, M., and Wilken, R. 2011. "Broadband in the burbs: NBN infrastructure, spectrum politics and the digital home", *M/C Journal* 14(4). Available online at http://journal.media-culture.org.au/index.php/mcjournal/article/view/400. Accessed March 3, 2018.

Barratt-Pugh, L., Bahn, S., and Gakere, E. 2013. "Managers as change agents", *Journal of Organizational Change Management* 26(4): 748–764.

Barrett, P. and McFarlane, M. 2002. "E-government and joined-up government". In *Global working group meeting*. Available online at www.anao.gov.au/sites/g/files/net616/f/Barr ett_e-government_and_joined%20up_government_2002.pdf. Accessed March 3, 2018.

Binning, C. and Young, M. 1999. *Beyond roads, rates and rubbish. Opportunities for local government to conserve native vegetation*. National R and D Program on Rehabilitation,

Management and Conservation of Remnant Vegetation, Research Report, 1/99, Environment Australia, Canberra.

Blackburn, G. 2014. "Elements of successful change: The Service Tasmania experience to public sector reform", *Australian Journal of Public Administration* 73(1): 103–114. DOI: 10.1111/1467-8500.12054.

Blackburn, G. 2016. "One-stop shopping for government services: Strengths and weaknesses of the Service Tasmania experience", *International Journal of Public Administration* 39(5): 1–11. DOI: 10.1080/01900692.2015.1015555.

Blainey, G. 1982. *The tyranny of distance: How distance shaped Australia's history.* Sydney, Australia: Pan Macmillan Australia.

Borins, S. 2003. *New information technology and the public sector in Ontario.* A Report to the Panel on the Role of Government, June. Available online at http://citeseerx.ist.psu.edu/viewdoc/download;jsessionid=DD996E818517F5739134C8E19D47A4D4?doi=10.1.1.467.5497andrep=rep1andtype=pdf. Accessed March 13, 2018.

Bourne, K., Nash, A., and Houghton, K. 2017. *Pillars of communities: Service delivery professionals in small Australian towns 1981–2011.* The Regional Australia Institute. Available online at www.regionalaustralia.org.au/home/wp-content/uploads/2017/12/RAI_Pillars-of-Communities_Small-Towns-Report.pdf. Accessed February 25, 2018.

Bovey, W.H. and Hede, A. 2001. "Resistance to organizational change: The role of defence mechanisms", *Journal of Managerial Psychology* 16(7): 534–548.

Boyd, D.P. 2011. "Lessons from turnaround leaders", *Strategy and Leadership* 39(3): 36–43.

Brand Tasmania Council Inc. 2014. *Brand Tasmania.* Available at www.brandtasmania.com. Accessed May 3, 2018.

Burton, R. 2014. *Service Tasmania annual report 2013–14.* Service Tasmania Unit, Department of Premier and Cabinet. Available online at www.dpac.tas.gov.au/__data/assets/pdf_file/0003/253263/Service_Tasmania_Annual_Report_2013-14.pdf. Accessed March 11, 2017.

Carter, E.C. 2008. "Successful change requires more than change management", *Journal for Quality and Participation* Spring 31(1): 20–23.

Corradini, F., Forastieri, L., Polzonetti, A., Riganelli, O., and Sergiacomi, A. 2018. *Shared services center for e-government policy.* Available online at https://arxiv.org/pdf/1802.07982. Accessed March 10, 2018.

Cresswell, A.M., Burke, G.B., and Pardo, T.A. 2006. *Advancing return on investment analysis for government IT: A public value framework.* Available online at www.ctg.albany.edu/publications/reports/advancing_roi/advancing_roi.pdf. Accessed March 9, 2018.

Croger Associates Pty Ltd, Stenning and Associates Pty Ltd, and Appleyard, G. 2003. *The TIGERS report.* Canberra: National Office for the Information Economy. Available online at www.finance.gov.au/agimo-archive/__data/assets/file/0014/12173/TIGERS_program_summary.pdf. Accessed March 3, 2018.

Day, A., Crown, S.N., and Ivany, M. 2017. "Organizational change and employee burnout: The moderating effects of support and job control", *Safety Science* 100: 4–12.

Department of Premier and Cabinet. 2009. *A social inclusion strategy for Tasmania: Preliminary response.* Hobart. Available online at www.dpac.tas.gov.au/__data/assets/word_doc/0017/111293/WEB_VERSION_-_Preliminary_Response_to_A_Social_Inclusion_Strategy_for_Tasmania.doc. Accessed March 3, 2018.

Dugdale, A., Daly, A., Papandrea, F., and Maley, M. 2018. "Connecting the dots: Accessing e-government". *Future challenges for e-government: Accessibility* Discussion Paper

No. 16. Available online at www.researchgate.net/publication/242511423_CONNEC TING_THE_DOTS_ACCESSING_E-GOVERNMENT. Accessed March 3, 2018.

Fraser, C. 2009. "E-government: The Canadian experience", *Dalhousie Journal of Inter-disciplinary Management* 5: 1–14.

Frost, N. 2005. *Professionalism, partnership and joined-up thinking: A research review of front-line working with children and families.* Available online at http://lx.iriss.org.uk/content/professionalism-partnership-and-joined-thinking-research-review-front-line-working-children-. Accessed March 10, 2018.

Georgalis, J., Samaratunge, R., Kimberley, N., and Lu, Y. 2015. "Change process character-istics and resistance to organizational change: The role of employee perceptions of justice", *Australian Journal of Management* 40(1): 89–113.

Grace, M. and Coventry, L. 2010. "The co-location of YP4 and Centrelink in Bendigo, Australia: An example of partnership in action", *Journal of Social Work* 10(2): 157–174.

Hall, G., Boddy, J., and Chenoweth, L. 2016. "An adventurous journey: Social workers guiding customer service workers on the welfare frontline", *Aotearoa New Zealand Social Work Review* 28(3): 26–37.

Halligan, J. and Wills, J. 2008. *The Centrelink experiment: Innovation in service delivery.* Canberra, Australia: ANU Press.

Inayatullah, S. 2011. "City futures in transformation: Emerging issues and case studies", *Futures* 43(7): 654–661.

Johanson, K., Kershaw, A., and Glow, H. 2014. "The advantage of proximity: The distinctive role of local government in cultural policy", *Australian Journal of Public Administration* 73(2): 218–234.

Kim, S. 2014. "Local electronic government leadership and innovation: South Korean experience", *Asia Pacific Journal of Public Administration* November 30(2): 165–192.

Kohlborn, T., Weiss, S., Poeppelbuss, J., Korthaus, A., and Fielt, E. 2010. *Online service delivery models: An international comparison in the public sector.* Proceedings of the 21st Australasian conference on information systems (ACIS 2010), December, pp. 1–3, Brisbane, Australia.

Lewis, L.K. 2011. *Organizational change: Creating change through strategic communica-tion.* 5th ed. West Sussex, UK: Wiley-Blackwell Publishing Ltd.

McCann, J. 2001. *Centrelink annual review 2000 – 01.* Australian Department of Human Services. Available online at www.humanservices.gov.au/sites/default/files/documents/2000-2001-centrelink-annual-review.docx. Accessed March 9, 2018.

Nam, T. and Pardo, T.A. 2013. *Identifying success factors and challenges of 311-driven service integration: A comparative case study of NYC311 and Philly311.* Proceedings of the 46th Hawaii international conference on system sciences–2013. Available online at www.ctg.albany.edu/publications/journals/hicss_2013_philly-nyc311/hicss_2013_philly-nyc311.pdf. Accessed March 9, 2018.

Norman, R. 2010. "The Centrelink experiment: Innovation in service delivery by John Halligan and Jules Wills", *Australian Journal of Public Administration* 69(1): 106–107.

Oreg, S. 2006. "Personality, context and resistance to organizational change", *European Journal of Work and Organizational Psychology* 15: 73–101.

Reddicka, C.G. and Turner, M. 2012. "Channel choice and public service delivery in Canada: Comparing e-government to traditional service delivery", *Government Information Quarterly* 29(1): 1–11.

Reid, R. and Wettenhall, R. 2015. "Shared services in Australia: Is it not time for some clarity?", *Asia Pacific Journal of Public Administration* 37(2): 102–114.

Rundle, T. 1997. Directions Statement – the statement of government initiatives delivered by the Honourable Mr Tony Rundle, T. MHA, Premier, Department of Premier and Cabinet, 10 April, Hobart, Australia.

Rundle, T. 2014. "Minority rule garnered milestones, says former Tasmanian premier Tony Rundle". *The Mercury*. Available online at www.themercury.com.au/news/opinion/minority-rule-garnered-milestones-says-former-tasmanian-premier-tony-rundle/news-story/bdfbe882800235e7f3a6467298811938. Accessed March 11, 2017.

Service New Brunswick. 2012. Available online at www.snb.ca. Accessed January 18, 2017.

Service Tasmania. 2018. Available online atwww.service.tas.gov.au/about/shops. Accessed August 15, 2018.

Shah, R. 2001. "One-stop-shop: Making government faster and friendlier", *PM: Public Management* 83(6): 16–19.

Sharma, V., Guttoo, D., and Ogra, A. 2014. *Next generation citizen centric e-services*. IST-Africa conference proceedings, Le Meridien Ile Maurice, Mauritius, May, pp. 1–15.

Sher, J.P. and Sher, K.R. 1994. "Beyond the conventional wisdom: Rural development as if Australia's rural people and communities really mattered", *Journal of Research in Rural Education* Spring 10(1): 2–43.

Sokvitne, L. 2002. "Aligning opportunities and technology, the challenge for libraries", *The Australian Library Journal* 51(2): 165–172. DOI: 10.1080/00049670.2002.10755986.

Szirom, T., Hyde, J., Lasater, Z., and Moore, C. 2001. "Working together–Integrated governance, beyond traditional boundaries". IPAA National Conference 2001, Sydney Convention Centre, November 28–30. Available online at http://citeseerx.ist.psu.edu/viewdoc/download?doi=10.1.1.195.3576andrep=rep1andtype=pdf. Accessed March 3, 2018.

Tasmanian Council of Social Service. 2009. *Just scraping by? Conversations with Tasmanians living on low incomes*. Sandy Bay: Tasmanian Council of Social Service. ISBN 978-0-9805301-2-4.

Vardon, S. 2002. "Centrelink, changing culture and expectations". In E.M. Milner (ed.), *Delivering the vision: Public services for the information society and the knowledge economy* (pp. 39–62). London, UK: Routledge.

Vincent, C. and Prychodk, N. 2002. "Working horizontally across the Canadian public sector", *Canadian Government Executive* 6: 18–20.

Ward, S., Lusoli, W., and Gibson, R. 2007. "Australian MPs and the Internet: Avoiding the digital age?", *Australian Journal of Public Administration* 66(2): 210–222.

4 Shared services in Australian local government

The case of the common service model

Brian Dollery

Introduction

In local government systems across the world, small regional, rural, and remote local authorities face far greater challenges than their metropolitan cousins, not only in terms of fiscal constraints, but in numerous other operational difficulties, not the least of which are problems in attracting staff with specialist administrative and technical skills (Bel and Warner, 2015; Lægreid *et al.*, 2016). Given the vast spatial scale involved, these problems are especially acute in regional, rural, and remote Australian local government, which confronts the ever-present 'tyranny of distance' (Dollery and Akimov, 2008b). While various methods have been advanced for tackling these challenges, resource-sharing, shared services, and other forms of inter-local cooperation between local councils represent the most promising approach (Dollery *et al.*, 2012). Furthermore, recent technological developments associated with web-based systems of regional collaboration and service provision have improved the delivery of back-office functions and local services in an efficient fashion over long distances. This chapter explores the most advanced model of shared service provision to small remote, rural, and regional local authorities in Australian local government in the form of the Common Service Model.

In Australian local government, this mode of shared service provision has been pioneered by the Brighton Council as its Common Service Provision Model (De Souza and Dollery, 2011). In contrast to most existing shared service platforms in Australian local government, the Common Service Model is wholly owned by the Brighton Council: it provides the same council functions and services to other local government entities as Brighton uses itself on a commercial 'fee-for-service' basis. The Common Service Model has not only proved to be extraordinarily successful in addressing the requirements of small non-metropolitan local authorities, but it has also been able to continuously improve the range of back-office functions and local services which it offers, as well as the nature of its service provision, by continuously adopting new technologies. This chapter provides an analysis of how the Common Service Model has been able to overcome many of the difficulties encountered by regional, rural, and remote Australian local authorities which spring from the characteristics of these local entities and the environmental circumstances in which they operate.

The chapter itself is divided into five main parts. Following this introduction, the second section provides a synoptic review of the literature on shared services in Australian local government. This is followed by a section which considers the nature of the challenges confronting remote, rural, and regional local government in Australia, whereas the subsequent section discusses the Australian national and state policy parameters within which local authorities must operate, especially as indicated in various state and national inquiries into local government. The next section examines the nature and operation of the Common Service Model. The chapter concludes with a brief assessment of the policy implications of the analysis, including its transferability to local government systems outside Australia.

Shared services in Australian local government

Oakerson (1999, p. 7) advanced the conceptual foundations for shared services in local government by developing a critical distinction between local service provision and local service production. Whereas local production creates a local service, by contrast local service provision comprises determining whether to provide a particular service, the quantity and quality of the service, and how it should be secured (Feiock, 2013). A substantial literature exists on the empirical analysis of shared services and the challenges it presents (see, for example, Bel and Warner, 2015; Carr and Hawkins, 2013; Henderson, 2015; Noda, 2017). The separation of local service provision from local service production enables local authorities to choose between different methods of providing services, such as the Common Service Model.

Numerous shared service arrangements exist in Australian local government, many revealing high levels of creativity and ingenuity, largely as a consequence of the spatial challenges which local councils confront. De Souza and Dollery (2011) have identified two main strands in the Australian literature on local government which has focused on the question of shared services. Firstly, a conceptual literature has addressed the theoretical aspects of shared services. For instance, Dollery *et al.* (2010) have classified shared service models in terms of various typological systems. Moreover, several scholars have attempted to describe the different institutional vehicles which can provide shared service provision, including Dollery and Johnson (2005) and the Local Government Association of Queensland (LGAQ) (2005).

Secondly, an empirical strand of the literature has investigated existing shared service models which have been employed in Australian local government, including Dollery and Akimov (2008a, 2008b, 2009) and Dollery *et al.* (2007a, 2012). This line of inquiry can in turn be subdivided into the examination of (a) specific models which have been implemented and (b) specific models which have been advanced as suitable candidates for implementation by local authorities. Category (a) incorporates scholarly inquiry into existing institutional arrangements, like regional organizations of councils (ROCs) (Dollery *et al.*, 2005b; Dollery *et al.*, 2007b; Marshall *et al.*, 2007; Sorensen *et al.*, 2007), the Walkerville model (Dollery and Byrnes, 2006), and New England Strategic

Alliance of Councils (NESAC) (Conway *et al.*, 2011; Dollery *et al.*, 2005a). Category (b) covers *ad hoc* resource sharing models (Ernst and Young, 1993), virtual local government (Allan, 2001, 2003), joint board models (Dollery and Johnson, 2007; Shires Association of NSW, 2004; Thornton, 1995), and the Gilgandra Co-operative Model (Dollery *et al.*, 2006b).

Characteristics of rural and regional local government in Australia

In common with local government systems in many other nations, Australian local government exhibits considerable diversity (Dollery *et al.*, 2006a). This diversity is especially marked between metropolitan municipalities and their counterparts in regional, rural, and remote areas (Department of Infrastructure and Regional Development, 2015). Whereas local council population size, as well as population density and population composition, typically varies considerably, differences extend far beyond demography (Dollery *et al.*, 2012). For instance, other socio-economic characteristics of non-metropolitan local authorities, particularly lower average household incomes, almost inevitably mean that local community preferences for local council services differ in both kind and degree from metropolitan municipalities. Thus, people outside metropolitan Australia usually prefer lower levels of service provision at commensurately lower levels of property taxes.

In addition, since a preponderance of voters dwell in cities, the political structure of most Australian state and territory jurisdictions reflects this reality in terms of political representation. One consequence of this resides in a widespread belief that people in regional, rural, and remote areas are largely invisible to public policy makers. As a result, the broader representative role of local government in advocating for non-metropolitan areas is seen as especially important outside the major cities (Grant and Drew, 2017). The net effect is that local council mayors and councillors in regional, rural, and remote local authorities are assessed by constituents on the efficacy of their advocacy with higher tiers of government and, accordingly, tend to place greater emphasis on it.

Other 'environmental' factors amplify these differences. For example, the spatial size of rural councils with agriculturally based economies, and the concomitant need for an extensive road network, typically mean that local authorities of this kind are obliged to expend a substantial proportion of their resources on road construction and maintenance. Furthermore, the low population densities representative of regional, rural, and remote Australia necessarily make many kinds of routine municipal service provision much more expensive than in metropolitan areas due to the cost characteristics of these services. For instance, domestic waste collection and disposal outlays rise when greater distances are involved. Along analogous lines, remote local councils with small population sizes must often act as 'government of last resort' by offering commercial services typically provided by the private sector in metropolitan municipalities, such as banking, undertaking amenities, and medical services (Dollery *et al.*, 2013).

In general, the demographic, economic, and social circumstances of local government in regional, rural, and remote Australia render local authorities in these areas *sui generis* from a public policy perspective. Put differently, policy instruments routinely employed in metropolitan local government are often rendered ineffective or even counterproductive outside major cities. For example, the heavy emphasis traditionally placed on structural reform through municipal mergers by Australian local government policy makers to tackle financial and other problems is singularly misplaced where the 'tyranny of distance' makes it practically impossible to forcibly amalgamate local councils with huge spatial areas (see, for example, Dollery *et al.*, 2012, for a detailed review of the empirical evidence on the outcomes of municipal mergers in Australian local government). These considerations have led local government practitioners and researchers alike to search for alternative policy instruments more suited to the unusual problems faced by regional, rural, and remote local councils in Australia, especially resource-sharing and shared services (Dollery and Akimov, 2008a, 2008b).

Various scholars of local government in Australia and elsewhere have argued that the unique characteristics of non-metropolitan local authorities necessarily imply that orthodox shared service models which have been successful in the context of metropolitan municipal service provision must be adjusted if they are to be efficacious in regional, rural, and remote settings (see, for example, Chen and Thurmaier, 2009; Garcea and LeSage, 2005; Lago-Penas and Martinez-Vazquez, 2013; Noda, 2017; Tomkinson, 2007). Significant literature also exists on the question of barriers to participation in shared services in local government (see, for example, Bel and Warner, 2015; Carr and Hawkins, 2013; Hawkins, 2009; Tomkinson, 2007; Warner and Hefetz, 2008). The thrust of empirical work in this area has been to demonstrate that there are common barriers across different local government systems in different countries which impede participation in shared service arrangements. In the context of regional, rural, and remote Australian local government, Dollery *et al.* (2012) have argued that two factors are especially significant and thus must be catered for in the construction of shared service models. Firstly, they contend that the design of shared service entities not only affects their operational efficacy but also contributes to their longevity as successful organizations. Four potentially contentious elements seem to be particularly crucial: (a) the ownership distribution of assets and liabilities of participating member councils, (b) the determination of voting rights, (c) the attribution of the cost burden of establishing the shared service entity and its continuing costs of operation, and (d) the distribution of surpluses and losses among member local authorities. Secondly, they have argued that the nature of shared service model membership can have a decisive bearing on its effectiveness. They observed that a common phenomenon in regional, rural, and remote shared service models is the so-called 'convoy problem', through which the progress of a shared service entity hinges on the behaviour of the least enthusiastic participating council in an analogous manner to maritime fleets which sail at the pace of the slowest ship. Dollery *et al.* (2012) argue that the convoy problem can be ameliorated in two main ways: ensuring that shared service participation is voluntary and enabling participating municipalities to 'pick and choose' between the shared service options they wish to secure.

It is thus evident that meeting the various requirements for a successful shared service provision in regional, rural, and remote Australian local government is problematic and far from straightforward (Dollery *et al.*, 2008; Dollery *et al.*, 2004; Kortt *et al.*, 2012). Indeed, the various impediments to efficacious resource sharing and shared services no doubt explain why so many attempts have failed over the years, such as the New England Strategic Alliance of Councils (NESAC) (Conway *et al.*, 2011). However, as we shall endeavour to demonstrate in this chapter, the Brighton Common Service Model has flourished in providing services to regional, rural, and remote councils in large part because it can accommodate all the requisite elements necessary for success.

Australian national and state policy parameters

With the sole exception of the Australian Capital Territory, all Australian states and territories have their own local government system. Australian local government provides a comparatively limited range of functions, mostly concentrated on 'services to property', including roads and drainage, waste collection, disposal, and recycling, and parks and recreational areas (Dollery *et al.*, 2006a). Unlike many local government systems in other developed countries, it does not offer many 'services to people', such as policing, public housing, and schools, which are state government responsibilities (Dollery and Robotti, 2008; Shah, 2006). However, in recent years local authorities have expanded service provision, which now often incorporates aged care, local museums and heritage sites, and land care programs (Dollery *et al.*, 2006c).

All Australian state and territory local government systems possess the legal parameters which allow local authorities to engage in resource sharing, shared services, and other forms of inter-municipal cooperative agreements (Dollery *et al.*, 2012). Indeed, over the past two decades in particular, steps have been taken to facilitate and encourage the adoption and expansion of shared services by most state governments. Much of the stimulation underpinning these developments arose as a result of national and state-based inquiries into local government which considered *inter alia* the role of shared services as a potential approach to addressing the problem of financial sustainability in local government.

In 2003, the (then) Commonwealth Government initiated an inquiry into local government which resulted in the publication of the Hawker Report (2003) entitled *Rates and Taxes*. It argued that the Commonwealth Government should foster "established Regional Organizations of Councils (ROCs) and other regional bodies which have demonstrated their capacity to be involved in the regional planning and delivery of federal and state government programs" (House of Representatives Standing Committee on Economics, Finance and Public Administration, 2003, p. 97). At the national level, the Hawker Report (2003) was followed by the PricewaterhouseCoopers (PwC) (2006) *National Financial Sustainability Study of Local Government* report. PwC (2006) found that shared services could substantially improve the operational efficiency of local authorities. In particular, PwC (2006, p. 121) argued that three aspects of shared service provision offered great promise: the emergence of specialized "lead service providers" in collaborating constellations of councils in which every participating local council developed expertise in a specific service and then provided this service to other member municipalities on a sound commercial basis; the advent of joint "bulk purchasing" and "procurement of goods and services" by groups of councils; and the greater use of combined "back-office" services, such as administration, finances, human resources, and information technology. PwC (2006, p. 121) contended that these avenues would generate "cost savings, productivity improvements and better training for staff".

The findings of these national inquiries on the value of shared services were echoed in a number of similar state-based investigations into local government financial sustainability. Thus, South Australian Financial Sustainability Review Board's (FSRB) (2005) *Rising to the Challenge* report, the New South Wales (NSW) Independent Inquiry into the Financial Sustainability of Local Government's (LGI) (2006) *Are Councils Sustainable?* report, the Local Government Association of Queensland's (LGAQ) (2005) *Size, Shape, and Sustainability Review Framework*, and the Western Australian Local Government Association's (WALGA) (2006) *Systemic Sustainability Study* all established that shared services could make a substantial contribution to financial sustainability in local government.

An important consequence of these national and state inquiries was to provide impetus to state governments to improve the legislative underpinnings for shared services. For example, after an unfortunate delay the NSW Government has finally launched legislation for the introduction of joint organizations (JOs) of councils which are empowered *inter alia* to offer extensive shared services. The proposed Local Government Amendment (Regional Joint Organizations) Bill 2017 offers the legal foundations necessary for JOs with shared service platforms to provide local functions and local services on a commercial basis to local authorities in NSW and elsewhere. Thus, for instance, schedule 1(10) (c) enables a JO to be "constituted as a body corporate with the powers of an individual both in and outside the State" and schedule 1(10) (e) allows JOs to "deliver services to or on behalf of councils and provide assistance to councils (including capacity building)". Analogous regulations exist or are planned in other state local government systems.

Common service model

The Brighton Council (see Figure 4.1) established its Common Service Model in 2007 in order to (a) generate additional income and (b) assist other small local authorities to operate more efficiently. In essence, the Common Service Model represents a resource-sharing arrangement which allows the Brighton Council to offer various services which it uses itself, such as financial management, IT, and planning, to other local councils. Over time, the Common Service Model has attracted numerous client municipalities, including Tasmanian councils such as Flinders Island, Glamorgan-Spring Bay, and Tasman; local authorities in continental Australian states and territories; and Suva City in Fiji (De Souza and Dollery, 2011).

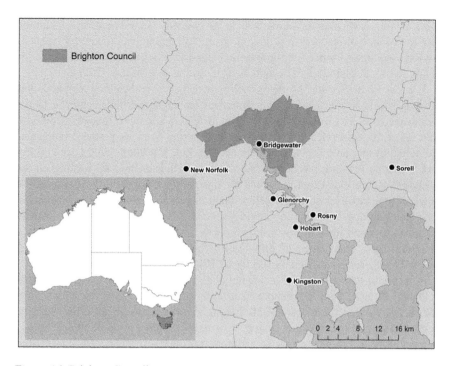

Figure 4.1 Brighton Council
Map credit: Aita Bezzola.

The Common Service Model has been deliberately designed to accommodate solutions to the problems faced by remote, rural, and regional councils. For example, almost all non-metropolitan councils experience great difficulties in recruiting professional and technically skilled personnel. Similarly, many of these municipalities struggle to keep on top of new regulations imposed by state departments of local government and other state and commonwealth agencies. These and other problems have spawned a growing reliance on commercial consultants, as well as greater resource sharing with other local authorities. However, both of these approaches have proved problematic: for-profit consultants are expensive and resource sharing has often not been successful (De Souza and Dollery, 2011).

These difficulties derive from the special circumstances of remote, rural, and regional councils. For instance, in terms of the need for commercial consultants, whereas large city councils are typically able to provide sophisticated services 'in-house', and if in-house capacity has limitations, then these municipalities can easily afford to employ consultants. However, these options are typically not available to small non-metropolitan councils, which cannot afford commercial consultants and do not have the requisite skills in-house. The Common Service

Model has been deliberately planned to meet this gap in skilled local government services for small and medium-sized local councils. In essence, the Common Service Model offers services ranging from full municipal functions, such as payroll and rates notices, to providing important regulatory skills, such as compliance and planning and functions (De Souza and Dollery, 2011).

The Common Service Model, wholly owned by the Brighton Council, has proved especially adept at keeping abreast of technological developments. Under the auspices of its CouncilWise platform, Brighton has more recently secured a suite of software services by forming a consortium with its partner companies, including leading providers such as Assetic Pty Ltd, Microsoft Ltd, Microwise Pty Ltd, Dialogue Pty Ltd, AvePoint Pty Ltd, and SaaSplaza Pty Ltd. Resting on the Microsoft Azure cloud and presented by means of Office 365, the CouncilWise range of municipal services now offers client councils state-of-the-art management tools and auditing functions to deliver financial, asset, and property management functions to small local authorities, irrespective of spatial impediments. It should be noted that a wide range of software platforms are available and we consider CouncilWise because it is employed in the Brighton Common Service Model.

In response to the needs of small regional, rural, and remote Australian local councils, CouncilWise has adapted software and internet systems which have substantially reduced the costs of IT. Through its platforms, it can offer property, rates, and animal management systems to council which are modified and refined to meet particular council needs. In addition, it provides financial systems suited to the business operations of the small councils. These contain numerous categories – property and rates, financial management, asset management, records management, customer relations management, permits and licencing, animal management, development and planning systems, communications and collaboration, and payroll and human resources – which can be tailored according to the requirements of specific client municipalities. A noteworthy feature of CouncilWise resides in its ability to incorporate new technologies as they become available. Costs are contained since the Brighton Common Service can leverage scale through its consortium arrangements with software and program suppliers not available to individual local authorities. This means it can readily be adapted, enhancing its transferability into other local government systems.

The operation of the basic Common Service Model is straightforward. The council providing the service in question offers the client council the service at an agreed price calibrated according to the type, term, and complexity of the service provided. De Souza and Dollery (2011, p. 6–7) interviewed senior executives at Brighton Council on the conditions which must be satisfied for the Common Service Model to operate successfully. These managers stressed several factors. For example, "provider and participating councils must enjoy 'existing friendly relations' at both elected member and staff levels". In addition, "there must be 'open and frank discussions' between councils preceding any agreement" and "neither participating councils nor the provider council should 'feel threatened' by each other". Furthermore, "experience has shown that the Brighton Model can begin on a small scale, growing one step at a time with 'confidence and maturity'", "services must be provided on commercial terms acceptable to both parties", "many services can be provided remotely with some regular 'on-site' presence", and "common platforms and systems should be used, such as the same computer software, for all participating councils" (De Souza and Dollery, 2011, p. 6–7).

Any successful shared service model designed to meet the needs of regional, rural, and remote local authorities must accommodate four key elements: it must minimize costs, maximize flexibility, ensure independent oversight, and embody voluntarism (Dollery *et al.*, 2016). Furthermore, the principle of "voluntary engagement" must be respected at all times and an appropriate organizational structure adopted. Incorporation of these elements will also make the model more easily transferable to other local government systems in other countries. This argument begs the question of how best a shared service model should be structured to incorporate these features.

In essence, a shared service model, like the Common Service Model, should be structured along the following lines (Dollery *et al.*, 2016, p. 236): Participating local authorities should "establish a Board made up of the General Managers of the interested councils, which would be chaired by an agreed Chair and supported by an agreed Chief Administrator". In turn, "the Board would establish a Common Service Model based on numerous *ad hoc* but formalized Shared Service Agreements (SSAs)", with a specific SSA for each defined service between a given provider council and a given recipient council. Every specific SSA would define precisely the nature and scope of the local service or local function in question as well as its costs. The implementation and subsequent operation of every SSA falling under the auspices of the shared service model would be monitored by the chair and chief administrator, who would be advised by a committee comprised of representatives from each of the provider and recipient councils involved. Dispute provisions, extensions, and other details of the SSA would be specified in every SSA.

Furthermore, this arrangement facilitates voluntary engagement, as the "voluntary participation by councils necessarily implies that the optimal [spatial] area for a given service is decided by participating municipalities using detailed local information" (Dollery *et al.*, 2016, p. 236). In addition, since each SSA only applies to participating councils, there are no issues concerning ownership structures and any operating surpluses and deficits "are not relevant to all member councils because they apply to participating councils only and have in any event already been agreed under the SSA" (p. 236).

SSAs established along these lines offer a number of advantages to shared service models such as the Common Service Model. In the first place, these SSAs enable participating councils to acquire "access to expertise in circumstances where councils cannot otherwise acquire these skills themselves, due to expense or the availability of appropriately skilled workers, a common problem faced by regional, rural, and remote local authorities" (Dollery *et al.*, 2016, p. 237). Secondly, it allows small local authorities "to fully deploy their existing staff and generate revenue from them, where they are 'under-employed'" (p. 237). Finally, it enables councils "to secure commercial returns on investments in IT systems, as well as other capital investments, by making these available to other local authorities" (p. 237).

Policy implications

As we have seen in this chapter, in common with their counterparts in other comparable countries, small regional, rural, and remote local authorities in all Australian state and territory local government systems face daunting problems, not the least are harsh ongoing financial constraints, as well as difficulties in securing administrative and technical skills. Since the 'tyranny of distance' renders structural solutions to these problems based on council consolidation impractical (Dollery *et al.*, 2012), we have argued that other policy remedies must be pursued, such as shared services and other forms of inter-municipal collaboration. However, given the problems faced by small non-metropolitan local authorities, which derive in large part from the peculiar characteristics of these councils and their environmental circumstances, careful consideration must be accorded to these characteristics in the design and implementation of appropriate shared service platforms.

Dollery *et al.* (2016) identified four main features which must be incorporated into any shared service model created to meet the requirements of regional, rural, and remote local councils: cost minimization, flexibility maximization, independent oversight, and voluntarism. This chapter has sought to demonstrate that a generic Common Service Model, designed along the lines of the existing Brighton Common Service Model, with its CouncilWise platform for delivering online services, best meets these four elements. Moreover, the ongoing success of the Brighton Common Service Model (De Souza and Dollery, 2011; Dollery *et al.*, 2016) attests to the fact that it has worked well in the institutional milieu of regional, rural, and remote Australian local government. Indeed, the cost-effective nature of the local functions and local services offered by the Common Service Model has enabled numerous small non-metropolitan municipalities to provide efficient and relevant modern services to local residents in an affordable manner. This would not have been possible without inter-council collaboration.

Since regional, rural, and remote local governments in many other developed federal countries face similar problems to their Australian state and territory cousins due to their common characteristics, a suitably adapted Common Service Model could easily be designed for American, Canadian, German, Japanese, New Zealand, and other similar multi-tiered governmental systems. In sum, the Common Service Model can readily be transferred to other local government systems outside Australia given its inherent capacity for flexibility.

References

Allan, P. 2001. *Secession: A manifesto for an independent Balmain Local Council.* Balmain: Balmain Secession Movement.

Allan, P. 2003. "Why smaller councils make sense", *Australian Journal of Public Administration* 62(3): 74–81.

Bel, G. and Warner, M.E. 2015. "Inter-municipal cooperation and costs: Expectations and evidence", *Public Administration* 93: 52–67.

Carr, J. and Hawkins, C.V. 2013. "The costs of cooperation: What the research tells us about managing the risks of service collaborations in the US", *State and Local Government Review* December 45: 224–239.

Chen, Y.-C. and Thurmaier, K. 2009. "Inter-local agreements as collaborations", *American Review of Public Administration* 39: 536–552.

Conway, M., Dollery, B.E., and Grant, B. 2011. "Shared service models in Australian local government: The fragmentation of the New England Strategic Alliance five years on", *Australian Geographer* 42(2): 207–223.

De Souza, S. and Dollery, B.E. 2011. "Shared services in Australian local government: The Brighton Common Service Model", *Journal of Economic and Social Policy* 14(2): Article 4.

Department of Infrastructure and Regional Development. 2015. *Local government national report, 2014/15*. Canberra: Department of Infrastructure and Regional Development.

Dollery, B., Grant, B., and Kortt, M.A. 2012. *Councils in cooperation: Shared services in Australian local government*. Sydney: Federation Press.

Dollery, B., Kortt, M.A., and Grant, B. 2013. *Funding the future: Financial sustainability and infrastructure finance in Australian local government*. Sydney: Federation Press.

Dollery, B.E. and Akimov, A. 2008a. "A critical comment on the analysis of shared services in the Queensland local government association's 'Size, Shape and Sustainability Program'", *Accounting, Accountability and Performance* 14(2): 29–44.

Dollery, B.E. and Akimov, A. 2008b. "Are shared services a panacea for Australian local government: A critical note on Australian and international empirical evidence", *International Review of Public Administration* 12(2): 1–11.

Dollery, B.E. and Akimov, A. 2009. "Shared services in Australian local government: Rationale, alternative models and empirical evidence", *Australian Journal of Public Administration* 68(2): 208–219.

Dollery, B.E., Akimov, A., and Byrnes, J. 2007a. *An analysis of shared local government services in Australia*. Working Paper 05-2007. Armidale: Centre for Local Government, University of New England.

Dollery, B.E., Burns, S., and Johnson, A. 2005a. "Structural reform in Australian local government: The Armidale Dumaresq-Guyra-Uralla-Walcha Strategic Alliance Model", *Sustaining Regions* 5(1): 5–13.

Dollery, B.E. and Byrnes, J. 2006. "Alternatives to amalgamation in Australian local government: The case of Walkerville", *Journal of Economic and Social Policy* 11(1): 1–20.

Dollery, B.E., Crase, L., and Johnson, A. 2006a. *Australian local government economics*. Sydney: University of New South Wales Press.

Dollery, B.E., Grant, B., and Akimov, A. 2010. "A typology of shared service provision in Australian local government", *Australian Geographer* 41(2): 217–231.

Dollery, B.E., Hallam, G., and Wallis, J.L. 2008. "Shared services in Australian local government: A case study of the Queensland Local Government Association Model", *Economic Papers* 27(4): 343–354.

Dollery, B.E. and Johnson, A. 2005. "Enhancing efficiency in Australian local government: An evaluation of alternative models of municipal governance", *Urban Policy and Research* 23(1): 73–86.

Dollery, B.E. and Johnson, A. 2007. "An analysis of the joint board or county model as the structural basis for effective Australian local government", *Australian Journal of Public Administration* 66(2): 198–207.

Dollery, B.E., Johnson, A., Marshall, N.A., and Witherby, A. 2005b. "ROCs governing frameworks for sustainable regional economic development: A case study", *Sustaining Regions* 4(3): 15–21.

Dollery, B.E., Kortt, M., and Drew, J. 2016. "Fostering shared services in local government: A common service model", *Australian Journal of Regional Studies* 22(2): 225–242.

Dollery, B.E., Marshall, N.A., Sancton, A., and Witherby, A. 2004. *Regional capacity building: How effective is REROC?* Wagga Wagga, NSW: Riverina Eastern Regional Organization of Councils.

Dollery, B.E., Marshall, N.A., and Sorensen, T. 2007b. "Doing the right thing: An evaluation of new models of local government service provision in regional New South Wales", *Rural Society* 17(1): 66–80.

Dollery, B.E., Moppett, W., and Crase, L. 2006b. "Spontaneous structural reform in Australian local government: The case of the Gilgandra cooperative model", *Australian Geographer* 37(3): 395–409.

Dollery, B.E. and Robotti, L. (eds.). 2008. *Theory and practice of local government reform.* Cheltenham: Edward Elgar.

Dollery, B.E., Wallis, J.L., and Allan, P. 2006c. "The debate that had to happen but never did: The changing role of Australian local government", *Australian Journal of Political Science* 41(4): 553–567.

Ernst and Young. 1993. *Resource sharing study: Town of St. Peters, Town of Walkerville, and City of Kensington and Norwood.* Canberra: Office of Local Government, Commonwealth Department of Housing and Regional Development.

Feiock, R.C. 2013. "The institutional collective action framework", *Policy Studies Journal* 41(3): 397–425.

Garcea, J. and LeSage, E. 2005. *Municipal reforms in Canada.* Toronto: Oxford University Press.

Grant, B. and Drew, J. 2017. *Local government in Australia: History, theory and public policy.* Amsterdam: Springer.

Hawkins, C.V. 2009. "Prospects for and barriers to local government joint ventures", *State and Local Government Review* 41(2): 108–119.

Henderson, A.C. (ed.). 2015. *Municipal shared services and consolidation.* London: Routledge.

House of Representatives Standing Committee on Economics, Finance and Public Administration ('Hawker Report'). 2003. *Rates and taxes: A fair share for responsible local government.* Canberra: Commonwealth of Australia.

Kortt, M., Dollery, B.E., and Grant, B. 2012. "Regional and local tensions: The role of shared services", *Public Policy* (Special Edition) 7(1): 33–43.

Lægreid, P., Sarapuu, K., Rykkja, L., and Randma-Liiv, T. (eds.). 2016. *Organizing for coordination in the public sector: Practices and lessons from 12 European countries.* Amsterdam: Springer.

Lago-Penas, S. and Martinez-Vazquez, J. (eds.). 2013. *Challenge of local government size.* Cheltenham: Edward Elgar Press.

Local Government Amendment Act (Regional Organizations) Bill 2017. Sydney: Legislative Assembly of New South Wales.

Local Government Association of Queensland (LGAQ). 2005. *Size, shape and sustainability of Queensland local government.* Discussion Paper. Brisbane: LGAQ.

Marshall, N.A., Dollery, B.E., and Sorensen, A. 2007. "Voluntary regional cooperation in Australia", *Canadian Journal of Regional Science* 29(2): 239–256.

New South Wales (NSW) Independent Inquiry into the Financial Sustainability of Local Government's (LGI). 2006. *Are councils sustainable?* Sydney: LGI.

Noda, Y. 2017. "Forms and effects of shared services: An assessment of local government arrangements in Japan", *Asia Pacific Journal of Public Administration* 39(1): 39–50.

Oakerson, R.J. 1999. *Governing local public economies*. Oakland: ICS Press.

PricewaterhouseCoopers (PwC). 2006. *National financial sustainability study of local government*. Sydney: PricewaterhouseCoopers.

Shah, A. (ed.). 2006. *Local governance in industrial countries*. Washington, DC: World Bank.

Shires Association of NSW. 2004. *Joint board model*. Sydney: Shires Association of NSW.

Sorensen, T., Marshall, N.A., and Dollery, B.E. 2007. "Changing governance of Australian regional development: Systems and effectiveness", *Space and Polity* 11(3): 297–315.

South Australian Financial Sustainability Review Board (FSRB). 2005. *Rising to the challenge final report*. Adelaide: FSRB.

Thornton, J. 1995. *The urban parish: An alternative approach to local government amalgamation*. Local Government Development Program Research Series, Office of Local Government, Canberra, Commonwealth Department of Housing and Regional Development.

Tomkinson, R. 2007. *Shared services in local government*. Aldershot: Gower Publishing.

Warner, M.E. and Hefetz, A. 2008. "Managing markets for public service: The role of mixed public–Private delivery of city services", *Public Administration Review* 68(1): 155–166.

Western Australian Local Government Association (WALGA). 2006. *Systemic sustainability study: In your hands - shaping the future of local government in Western Australia*. Perth: WALGA.

Part III

New service arrangements

5 Rural health service delivery challenges in an era of neoliberalism in New Zealand

Etienne Nel and Sean Connelly

Introduction

Since the 1980s, New Zealand, in parallel with many other countries, witnessed a clear shift in economic policy in response to global economic change and the perceived limitations of Keynesian interventions, embracing more market-based and supply-side solutions to economic and spatial planning (Mudge, 2008; Peck, 2013). New Zealand, however, pursued these changes far more aggressively than other countries, effectively removing all state subsidies, trade protection, and regional support, and began corporatizing state services and introducing a managerial approach to administration (Peet, 2012; Shone and Ali Memon, 2008). This transformed the country from a champion of the welfare state to one of the least interventionist countries in the world (Challies and Murray, 2008; Conradson and Pawson, 1997, 2009). According to Peet (2012, p. 151), "New Zealand is a particularly interesting case because of its well-deserved reputation as a social democratic, welfare state that went Neoliberal with a vengeance in the mid-1980s".

Most previous forms of spatial intervention and community economic support were axed in the 1980s and 1990s, which contrasts with most other western democracies where they have often persisted, albeit in diluted forms. Regions and small urban centres have had to rely on the vagaries of the market for their economic well-being, with the limited capacity of local government obliging many local communities to look inward to their own resources and capacities to ensure the future of their communities (Britton *et al.*, 1992; Haggerty *et al.*, 2008; Larner and Walters, 2000; Nel, 2015). The economic changes led, at one level, to the effective axing of once-significant levels of regional and rural support, while at another level to the rationalization of remaining state services and operations such as the railways, timber industry, hospitals, and postal services in particular, weakening the economic and social well-being of many small towns (Barnett and Barnett, 2003; Wilson, 1995). Loss of state support for local-level development and limited local government finances meant that civil society is obliged to play a greater role in terms of self-help (North, 2002).

One sector which was particularly hard-hit was that of health care, particularly in rural areas where declining population numbers and rising costs, in an era of austerity and 'new managerialism' (Lynch, 2014), have seen the application of market-based principles, particularly to rural health provision, where efforts to reduce costs and rationalize services have shifted burdens onto communities. Service provision shifted to an approach of 'new managerialism' under neoliberalism, with the state seeking to offload the costs of services such as health care to the individual. As a net result, "the viability of health services in rural communities has been threatened by a number of factors, including declining levels of population and economic activity, [and] the desire to restrain public sector expenditure on health overall" (Barnett and Barnett, 2003, p. 59). This has been exacerbated by rural disadvantage and rural skills shortages, the commodification of health care, and new centralized governance arrangements. The net result, which occurred in some countries, and as will be discussed below in relation to New Zealand, is that community health trusts have emerged as part of a new innovation for the co-production and co-delivery of public health services (Bovaird, 2007).

This chapter provides insight, through local case studies, into the impact of the rationalization of health care in rural and small town New Zealand under neoliberalism and how, and with what degree of success, selected rural communities, through the formation of local community health trusts, have been able to respond to both the continued needs of their communities and the reduced level of state care. After overviewing relevant literature on rural health care and community responses, the chapter examines three case study communities in the Otago region to discern and interpret the nature of community-based responses to rural health care changes.

The changing rural health care context

Rural health care restructuring

Neoliberalism has obligated rural and small town communities in New Zealand to become more proactive in terms of ensuring that local development occurs and that services continue to be provided, reflecting a dependence on new governance, local leadership, and community cohesion (Nel, 2012). This parallels findings on the 'new rural economy' in Canada regarding the importance of community, human, and economic links for communities facing change and the importance of community capacity, social cohesion, and social capital in the context of the loss of state support and services. The effectiveness of the response depends on the ability of the community to mobilize effectively, drawing on local assets and the strengths present in place (Sullivan *et al.*, 2014).

In the midst of the restructuring of global economies and public expenditure, Cordes (1989) identified how rural health services in the global North were shaped by the relationship between two external forces. The first force derives from changes in the provision of health services that have become more specialized, more reliant on technology, and more costly. At the same time, other forces have been underway that have restructured the social, economic, and demographic structure of rural places. The relationship between these two forces presents significant challenges for how rural health services are funded, delivered, and rationalized, and have continued into the present. The focus on the connection between rural health and place has oscillated between a focus on health issues facing populations defined by demographics and health issues in populations defined by their geographic location (Phillips and McLeroy, 2004).

Access to health services is also a major rural health issue. This need is often expressed through community efforts to recruit or retain a doctor or to preserve existing medical services. However, the trend in service provision is to reduce funding, infrastructure, and support for rural health services in efforts to centralize services in larger centres to achieve greater efficiencies. However, this does not match people's preferences to be cared for in their local environment (Strasser, 2003). This preference, often expressed colloquially as preserving the "ability to be born and to die" in rural places, explains the willingness of rural residents to mobilize and work together to preserve and protect existing services.

Reflecting on the impacts of this shift towards a neoliberal philosophy to guide government distribution of many health and welfare services in Australia, Alston (2007) outlines significant impacts on increasing social exclusion and declining access to services in rural places. These policies have widened the gap in health inequalities between rural and urban Australians (Alston, 2007). It is a similar story in New Zealand, where the ongoing restructuring of the health sector (and the public sector more widely) has reinforced the linkages between people, health, and place and has led many rural communities to see prospects for community development being closely tied to health care provision (Kearns and Joseph, 1997). However, the ability to successfully mobilize residents to protect and maintain services is uneven and dependent on pre-existing community leadership capability and the prominent role of local professionals (Barnett and Barnett, 2003).

Rural health or community development?

Equitable access and availability of rural health services is a persistent issue for rural people and places (Halseth and Ryser, 2006). With cuts to funding and services, rural places and communities have responded by redefining public health services by linking them directly to community development through the involvement of a wider range of actors beyond health practitioners in the provision of services. As a result, in many rural places, rural health provision has shifted from a top-down service delivery model driven by experts to include health system users and communities in the co-production and co-delivery of public health services (Bovaird, 2007). This collaborative approach to health service delivery involves communities developing expertise in their own health needs at a community level, with the role of health care professionals shifting from being fixers to facilitators (Realpe and Wallace, 2010). Such a shift requires a relocation of power towards service users in determining what services are provided and how they are provided so that they best meet the needs of the community. The innovative community health trust model in New Zealand provides a mechanism for communities to regain some control over rural health services. However, while greater community empowerment in rural health delivery is meant to be more locally responsive, be more cost effective, and result in improved health outcomes, policy approaches that enable greater community participation often fail to recognize that meaningful participation is both uneven and difficult (Kenny *et al.*, 2015). Greater reliance on unfunded work and voluntarism has exacerbated uneven access to health services between those communities that are able to draw on larger pools of volunteers and those that are not (Milligan and Conradson, 2006; Skinner *et al.*, 2014), particularly in ageing communities (Hanlon and Halseth, 2005).

Community participation, rural health partnerships, and social infrastructure

The context for rural health service delivery has changed rapidly, with much of the burden falling to communities to maintain and preserve services. Innovative responses to this challenge are characterized by increased community participation and partnership models that together contribute to the social infrastructure required to take risks, mobilize resources, and strengthen a sense of place.

Successful rural health service models in Australia and New Zealand are characterized by the intersection of a macro-level supportive health policy, local and regional governance mechanisms, and local community capacity that bring together social capital and leadership of individuals from both the health service and community sectors (Barnett and Barnett, 2003; Johns *et al.*, 2007; Humphreys *et al.*, 2008). Community leadership and capability and the role of rural health service providers as social entrepreneurs (Farmer and Kilpatrick, 2009) are critical in generating genuine community participation that is able to overcome challenges of limited funding and to draw on the wider resource base in the community that might be referred to as the social infrastructure of place.

Social infrastructure refers to investments of time and energy into facilities and organizations that foster the ability of people to come together to generate social capital and a collective sense of purpose (Seyfang and Haxeltine, 2012). It is the formal and informal governance structures and relationships that exist between people that create the political and social space where social capital can be activated to advance specific projects (Connelly and Beckie, 2016). These inter-active ties that link organizations, institutions, and individuals are crucial for generating diverse resources and community capacity to respond to agreed-upon goals (Flora and Flora, 1993).

As Seyfang and Haxeltine (2012) suggest, local projects require high levels of trust and engagement among participants. This engagement is often mobilized through the use of symbols and threats that create a shared sense of identity about a place that serves to motivate action around the need to respond to challenges and threats from elsewhere. The closure and/or the threat of closure of hospitals and health services are a particularly strong symbol around which communities can stake their identity. This shared sense of identity transcends both individual and community scales and contributes to the social infrastructure that allows participants to draw on individual and collective resources to respond to threats posed by the loss of services and generates unique synergies with other community development projects. In the case studies that follow, we highlight how the selective development of community health trusts in rural New Zealand reflects the availability of social infrastructure components of capacity, participation, and shared sense of place.

Co-production and co-delivery of rural health in the Otago region, New Zealand: Community health trust model

General

Historically, rural hospitals in New Zealand were locally funded up to the post-World War II era. In the 1960s, the state took over these services as part of the broader state welfare approach which prevailed at the time, but then, in 1980s and 1990s, it started to rationalize service provision in terms of the then recently adopted new managerialism approach. This was in part driven by a desire to reduce costs and rationalize administration, with a shift to a population-based funding model disadvantaging many rural communities in favour of urban ones (Barnett and Barnett, 2003) This led, inevitably, to moves to reduce the number of rural hospitals. In many rural communities, the impending loss of hospital access was a significant blow, leading many communities, from the 1990s, to start to form their own community health trusts in an effort to retain key services locally (Eyre and Gauld, 2003; Webb, 2013) and, by implication, retain the services needed to prevent future population loss. The response was community driven, and was initially expressed through protests. These proved ineffective and the response gave way, in many areas, to the co-provision of health care with community health trusts emerging to ensure continued service delivery, albeit with a degree of state financial subsidization. The trusts developed in response to the government allowing communities to form community companies to take over the provision of health care services from the state. While they vary in structure and focus, broadly speaking they are community-based corporations, representing local communities organized around largely voluntary activity to retain and support health provision in their locales (Bidwell, 2001; Eyre and Gauld, 2003).

In their overview of numerous trusts across the country, Bidwell (2001) and Barnett and Barnett (2003) note that they have enjoyed varying levels of success, depending for their survival on the strength of local leadership and innovation, their ability to lobby external support, local capability, the involvement of local professionals, having high degrees of voluntary and financial support, commitment, networking, a degree of continued state support, and having a broad base of accountability and legitimacy. They found that where trusts succeeded, they led to greater levels of community control and empowerment, improved access to services, and more sustainable provision of services. In other less positive cases, however, they found that the administrative burden was passed on to communities, the financial impact was significant, and some communities were polarized over the focus of trust activities.

In the area now covered by the Southern District Health Board (see Figure 5.1), which incorporates the two administrative regions of Southland and Otago, numerous small, previously state-funded hospitals were either closed or handed over to trusts from the 1990s, leaving the only directly state-run and funded hospitals in the two cities of Dunedin and Invercargill and the town of Queenstown (which has a small rural hospital facility only). Twelve community trusts emerged in this combined region, three of which serve as case studies below. These twelve trusts include those in the larger centres of Gore, Oamaru, and Clyde (including neighbouring Alexandra and Cromwell) and those in nine smaller towns – Winton, Tuatapere, Lumsden, Tapanui, Ranfurly, Lawrence, Milton, Roxburgh, and Balclutha (see Figure 5.1) (Key informant #1, 2017).

In the case of the smaller, latter nine towns, all had functioning independent hospitals, providing a combined total of 195 beds prior to 1993. After the reforms and the takeover by community health trusts, this fell steeply to six operational facilities and a combined total of 65 beds (Barnett and Barnett, 2003). Of the nine trusts, eight (since 2016) provide general practitioner and district nursing services (White, 2016a). Two of the nine trusts have no hospital bed facilities, while three of the remainder only provide maternity beds and the

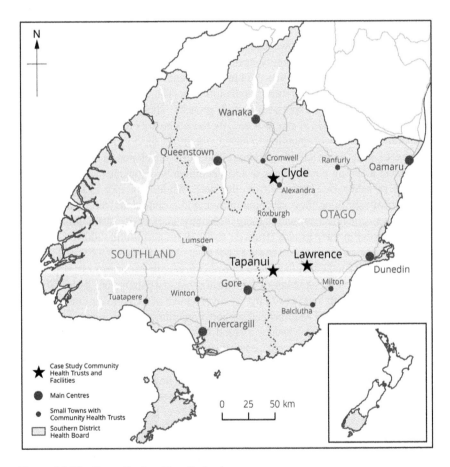

Figure 5.1 The Otago Region, New Zealand

remaining four have a combination of medical, maternity, and geriatric beds. The Southern District Health Board still technically owns the hospitals while overall management and provision of facilities is the responsibility of the trusts, which normally receive a patient-based subsidy for the provision of care from the Health Board (Key informant #1, 2017). Only in Lawrence, Roxburgh (in this case the state hospital was converted into an independent old age home), and Tapanui are the facilities directly owned by a community company.

In the next section, attention shifts to an examination of these three very different community health trusts to detail how these three communities responded, and with what degree of success, to the realities of neoliberalism and the need to co-provide health care in collaboration with the state if they wished to retain local health services. The first case, that of Lawrence, was the first community health trust in the country, and its set-up involved the direct takeover of a former state facility with the trust entering into a service agreement with the state, which provides a state subsidy for hospitalized patient expenses. In the second instance, that of Dunstan hospital in Clyde, the state continues to own the building and has a service agreement with the trust, which provides management and services. In the third case, that of Tapanui, the old hospital was closed and eventually demolished, and it was some 24 years before the local trust opened a new health and aged care facility, largely through the actions of dedicated community volunteers and local government support. The latter, not having hospital beds, does not receive the patient Health Board subsidy which the other two trusts receive. Both Lawrence and Tapanui illustrate cases of significant community-based, pro-active endeavours to largely self-fund essential services (with state support for hospital beds in Lawrence). In the case of Tapanui, local government support is interesting and shows how, in an era of state rationalism, a responsive local government has recognized its role in addressing shortfalls in the broader level of state health care to communities. The Dunstan example, following significant community action, saw the state essentially entering into a partnership model with the community to continue to provide needed services, while pursuing a new managerialist logic which sees the management of the facility and the provision of facilities being in the hands of the community.

Lawrence

The population of the small town of Lawrence (see Figure 5.1) peaked at some 800 people in the 1970s, but after that it started haemorrhaging residents as services were gradually rationalized with the introduction of neoliberalism and the declining productivity on the farms and of the forestry sector in the rural hinterland (Key informant #2, 2017). Such a small population base could not justify the continued operation of the small state hospital and when plans for its closure were made apparent, it immediately galvanized community action from local people desirous to retain the services and, in so doing, potentially prevent the further loss of residents (Key informant #3, 2012). The town had the benefit of numerous local sporting and service organizations, a community company which runs promotional activities, and a committed base of farmers in the district who rally support and raise funds (Key informant #2, 2017).

The first action by the community was to set up a community health trust, the first in the country, to take over health care provision from the state in terms of recent legislative changes (Bidwell, 2001; Eyre and Gauld, 2003). Initial funds to set up the trust were provided from proceeds raised by the local golf club (Key informant #3, 2017). Subsequently, the trust established a membership base with subscription being used to fund running costs and provide members with access to local health care. In 1992, the Tuapeka Health Company was formally set up to run the adjacent rest home and medical centres (Key informant #4, 2017). The hospital was acquired for the payment of a notional $1 (NZ) to the state, but the company had to immediately engage in fundraising to secure $1 million (NZ) to upgrade the facility. This was achieved through the support of 31 local groups, societies, and clubs (Eyre and Gauld, 2003). This led to the formation of a community-run, but effectively privatized, facility which provides 17 rest home beds, five health-board-subsidized hospital beds, general practitioner services (which also nets a state subsidy), community nursing, and meals on wheels in the community. When the local pharmacy closed and no pharmacist showed interest in taking over the business, the company bought the business, which it now leases to an out-of-town pharmacist who provides occasional personal services, but otherwise issues electronic prescriptions for local collection from his business in Roxburgh (Key informant #4, 2017).

The impact of retaining the facility locally is naturally significant. It provides continued medical care in the community, and scope for people to retire locally or be treated locally when in need of hospital-level care. Over and above the value of locally available services, a very direct impact is seen in the 23 local jobs which were retained or created through the community action. This includes one doctor, three nurses, the practice manager, district nurses, cleaners, and administrative staff. The annual budget is $1.1 million (NZ), of which the salary contribution to the local community is $0.7 million (NZ) (Key informant #4, 2017).

What has been achieved in Lawrence is clearly significant. A small rural town and its community have retained and extended multifaceted health care in their community for over a quarter of a century, responding to state rationalization and new managerialism through proactive local action. The company's future at this juncture is, however, not assured. The subsidy it receives from the Southern District Health Board is for the five hospital beds. The fact that the Health Board is facing further cuts to its budget in 2018 does not bode well for Lawrence, which may see this subsidy compromised. Aggravating the situation has been the falling population of the town, which in 2013 stood at 414 people (StatsNZ, 2018), the recent closure of a major employer, and the limits to the capacity of local organizations to continually support local facilities. The community recently built a new community pool, which some informants feel will limit future opportunities for local generosity, as local people may be reluctant to contribute once again to a major community initiative (Key informant #3, 2017;

Key informant #4, 2017). In terms of population, the local funding model suggests that the facility needs a potential pool of 1,400 people to remain financially viable, but the local population in the catchment area has fallen to 1,030. A recently commissioned consultant's report came up with three options: close, only provide medical services, or expand. The community favours the latter, but to do so they will need to raise another $2 million (NZ) for new facilities and increase the number of subsidized beds to 15, if the Health Board agrees (Key informant #4, 2017).

The Lawrence case is an inspiring one, showing the capacity of local residents to self-provide when the state rationalizes its activities. However, it also reveals the long-term risks when the population base is declining, subsidy options are limited, and the need for ongoing maintenance takes its toll, which in turn may compromise community capacity and responsiveness (Eyre and Gauld, 2003).

Clyde – Dunstan Hospital

The above-mentioned restructuring of health care in the rural areas of the country promoted local concern over the future of the Cromwell hospital and the Dunstan hospital situated in Clyde, but lying between the two larger settlements of Cromwell and Alexandra (see Figure 5.1). In anticipation of change, the municipality established the Dunstan Health Services Steering Committee to ensure services were retained at Dunstan Hospital. In 1990, the government decided to close the Cromwell facility, which was taken over by the Methodist Mission and is run as a hospital and rest home for the elderly (Cloete, 2013). Dunstan Hospital was run by the state through the 1990s, but when, in 1998, it became apparent that the government was withdrawing funds for the local hospital's management, the local community established the Central Otago Advisory Committee on Health, involving local doctors and community leaders. This committee identified the need to provide integrated health care in the district at both the primary and secondary levels, and raised local funds to develop a business plan. In 1998, the committee established Central Otago Health Incorporated (as a trust) which, in 1999, signed a contract with the Otago District Health Board (which preceded the Southern District Health Board) according to which the "community took [over] responsibility for the delivery of health services from Dunstan hospital" (Webb, 2013, p. 17), taking over direct management of the facility, and renegotiating service agreements. Unlike the Lawrence case, the hospital remains the property of the state and, given the larger size of the facility, a larger hospital patient bed (20 in total) subsidy is received, enabling the facility to offer a wider range of primary and secondary care locally and to the wider district (Key informant #1, 2017).

The well-organized local committee, community buy-in, concerns over the possible loss of health services, and the support of local health care professionals were clearly instrumental in ensuring the smooth transition of the facility to the community and continued provision of services through a co-funding/co-management arrangement (Flannery, 2013, p. 71). While Central Otago Health Services Ltd. is a community-owned, not-for-profit company which provides health services out of

Dunstan Hospital and owns the assets and facilities, a parallel community structure, Friends of Dunstan, established in 1992 over concern about the state of facilities in the hospital, has raised funds over the last 25 years to purchase new equipment and upgrade the facilities at the hospital (Flannery, 2013).

In 2003, the Otago District Health Board agreed to allocate $10 million (NZ) to new hospital buildings at Dunstan, on the condition that the local community raise the funds to pay for furniture, fittings, and equipment. As a direct response, Friends of Dunstan, with support from the community, service organizations, the local grant funding body, and businesses, raised the community's $2 million (NZ) co-payment for the new facility, which opened in 2005 (Flannery, 2013, p. 71). While this model does reflect an interesting compromise in the provision of rural health care, the co-management scenario does beg the questions of why rural residents have to pay more for health services than urban dwellers, and whether a poorer and less capacitated community would have been able to achieve the same.

The maintenance of the facility clearly provides for the continued, long-term health needs of the community. It is equally significant, from an economic point of view, that the hospital (and the associated services it provides) employs 130 staff and contractors (COHSL, 2017). Moving into the future, there is significant local concern about both the growing and the ageing populations. Between 2006 and 2013, the population of Clyde rose 9.8%, that of Cromwell by 15.6%, and that of Alexandra fell marginally (StatsNZ, 2018). Recent population growth is a result of growing tourism and retirement opportunities. These population and associated economic gains need to be balanced against the very real pressure growth they will put on the facility in the next 30 years, which will require continued local-level engagement and support to ensure that future health needs can be addressed (Key informant #1, 2017; Visser, 2013).

Tapanui

The experience of Tapanui (see Figure 5.1) shares similarities with that of Lawrence. In 1994, after 82 years of service, the local hospital was closed by the state following the rationalization of hospital services in the Otago-Southland region. As noted by White (2016b), as is the case in many communities, population loss, rising costs, and reduced funding led to closure of the local hospital. However, unlike Lawrence, where a community health trust was in place to take over the facility, the same was not the case in Tapanui. In response to the health-related challenges which emerged over time, posed by the absence of comprehensive health care provision and the growing needs of the ageing local population for retirement home facilities, the West Otago Health Trust was established, as a non-profit organization, in the town in 2002 to investigate options for a replacement facility (de Reus, 2014; MacLean, 2015). Additional motivations included the closure of local banks and other services which, in the community's view, was a serious deterrent to attracting new residents and a risk to retaining the existing population (stuff.co.nz, 19 August 2017).

In 2008, the Otago District Health Board (the predecessor of the Southern District Health Board) agreed to allow the centre to proceed and that it could provide age-related residential-care services. At the same time, the government gifted the land of the former hospital site to the trust, which provided the catalyst to the community to plan the facility's construction and to commence a long-term fundraising drive (Conway, 2008). Unlike the case in Lawrence, where the trust was able to repurpose the existing hospital, in Tapanui the former hospital was no longer fit for purpose and was demolished. The need of the community to provide a brand new facility forced them into a long-term fundraising commitment, with associated risks to the efficacy of the project. An eight-year funding drive was undertaken by the community, successfully raising $3.68 million (NZ), which paid for the 14-bed Ribbonwood Rest Home and the 24-hour-access West Otago Health Centre. The combined centres, now run by West Otago Health Ltd., employ 31 staff, making it the largest employer in the district, and serve the needs of its 2,000 residents (de Reus, 2013; White, 2016a, 2016b).

The community raised the money through 80 different fundraising events, including book and livestock sales, garden tours, concerts, local auctions, and significant local donations, which combined raised some $2 million (NZ); $1 million (NZ) was provided by the District Council, which will be repaid through a local rates levy; and the Dependent Cottages Fund provided $0.68 million (NZ) to assist with the costs of the rest home (MacLean, 2014; White, 2016a). In addition, 26,000 hours of volunteer time were needed to ensure the project's success. The facility which has been created stands as a testimony to what a motivated and unified community can achieve when it is obliged to look after its own future interests. According to Marianne Parks, the West Otago Health Ltd. director (in scoop.co.nz, 2015), "it's a project that galvanised our community, and is a living example of the magic that can happen when a group of committed individuals come together with a shared resolve, to make things happen". In the view of the district mayor, Bryan Cardogan, what has been achieved is "a reward for a community that works well together, that gets stuck in and looks after its own" (MacLean, 2014). The local success was recognized nationally when in 2016 the Health Trust won the Trustpower National Community Award for being a "great example of a community group taking ownership of an opportunity and developing it further" (nzDoctor.co.nz 27 April 2017).

Assessment and conclusion

The three cases of community engagement in rural health care examined above reflect the degree to which the New Zealand state's embrace of neoliberalism catalyzed the gradual withdrawal of state rural health services on cost grounds and the passing on of the responsibility to provide such services to those local communities motivated and organized enough to accept and respond to new co-management opportunities (Barnett and Barnett, 2003; Lynch, 2014; Peet, 2012). There has been a reliance on community fundraising, voluntarism, and a redefinition of health facilities in response to community mobilization around the threat of lost health services.

The three cases parallel findings from Canada where Sullivan *et al.* (2014) identified the importance of community capacity, social cohesion, and social capital in responding to the needs of communities undergoing transition. It is, however, important to note that in New Zealand, while "some communities have been successful in maintaining local access to a range of services . . . these success levels have been inconsistent" (Bidwell, 2001, p. 44). The successful cases of community responses examined in this chapter bear out the key catalytic factors noted by Bidwell (2001) and Barnett and Barnett (2003), including the importance of strong leadership, committed local professionals, networking, local dedication, and support from affected communities. They also show the importance of social infrastructure in fostering social capital and trust to assist in local mobilization (Seyfang and Haxeltine, 2012).

The cases show the importance of fostering linkages between residents, health, and place (Kearns and Joseph, 1997); the importance of voluntarism (Skinner *et al.*, 2014); and what the co-production and co-delivery of public health services look like in rural areas (Bovaird, 2007). Moving forward, however, community health trusts face significant long-term challenges. These include maintaining what may now be ageing facilities, ensuring continued community buy-in, particularly in areas where populations may be declining and those who remain are ageing, and the risk of future cuts to the very limited state subsidies which remain. The latter considerations face Lawrence at this juncture and may well face Tapanui in the future. While Dunstan was fortunate to retain a higher level of state co-funding, its challenge will be that of coping with very real population growth and the pressure that is placing on the facility.

The three cases reflect different degrees of community action and co-management of rural health in an era of neoliberalism. Lawrence involved direct community takeover; Tapanui was based on the same model but the involvement of the local authority provides an interesting example of local state response to the withdrawal of central state services. Unlike these two cases, in Dunstan, the population was not declining and the existence of several large settlements near the hospital meant that the District Health Board retained full control for longer and continues to have a high degree of involvement with the facility. This does raise the questions of why the state withdrew services in the first place from Dunstan and to what degree the state sought to pass on many of the costs of health care to a community whose size and growth should have justified the continued provision of state care. It is apparent that the funding model and approach have negatively impacted rural areas and the small towns which they host, devolving the management and responsibility of health care to rural communities in the context of reduced state financial input. The urban bias implicit in this scenario serves to reinforce geographically uneven development, which is the spatial hallmark of neoliberalism (Harvey, 2015).

References

Alston, M. 2007. "Globalisation, rural restructuring and health service delivery in Australia: Policy failure and the role of social work?", *Health & Social Care in the Community* 15(3): 195–202. doi: 10.1111/j.1365-2524.2007.00696.x.

Barnett, R. and Barnett, P. 2003. "'If you want to sit on your butts you'll get nothing!' Community activism in response to threats of rural hospital closure in southern New Zealand", *Health and Place* 9: 59–71.

Bidwell, S. 2001. *Successful models of rural health service delivery and community involvement in rural health: International literature review.* Christchurch: Centre for Rural Health.

Bovaird, T. 2007. "Beyond engagement and participation: User and community coproduction of public services", *Public Administration Review* 67(3): 846–860.

Britton, S., Le Heron, R., & Pawson, E. 1992. *Changing Places in New Zealand: A geography of restructuring.* Christchurch: New Zealand Geographical Society.

Challies, R.T. and Murray, W.E. 2008. "Towards post-neoliberalism? The comparative politico-economic transition in New Zealand and Chile", *Asia Pacific Viewpoint* 49(2): 228–243.

Cloete, J. 2013. "The Cromwell Hospital". In Webb, A. (ed.). *Dunstan hospital clyde* (pp. 98–101). Cromwell: Central Otago Health Inc.

COHSL (Central Otago Health Services Limited). 2017. www.cohsl.co.nz/about-us. Accessed 8/8/2017.

Connelly, S. and Beckie, M. 2016. "The dilemma of scaling up local food initiatives: Is social infrastructure the essential ingredient?", *Canadian Food Studies* 3(2): 49–69.

Conradson, D. & Pawson, E. 1997. "Reworking the geography of the long boom: The small town experience of restructuring in Reefton, New Zealand", *Environment and Planning A* 29: 1381–1397.

Conradson, D. and Pawson, E. 2009. "New cultural economies of marginality: Revisiting the west coast, South Island, New Zealand", *Journal of Rural Studies* 25(1): 77–86.

Conway, G. 2008. Health facility given go ahead. *Otago Daily Times* May 28. Available online at www.odt.co.nz/regions/south-otago/health-facility-given-go-ahead. Accessed 19/8/2017.

Cordes, S.M. 1989. "The changing rural environment and the relationship between health services and rural development", *Health Services Research* 23(6): 757–784.

De Reus, H. 2013. Health centre progressing. *Otago Daily Times* November 8. Available online at www.odt.co.nz/regions/south-otago/health-centre-progressing. Accessed 19/8/2017.

De Reus, H. 2014. Building of health centre on schedule. *Otago Daily Times* January 21. Available online at www.odt.co.nz/regions/west-coast/building-health-centre-schedule. Accessed 19/8/2017.

Eyre, R. and Gauld, R. 2003. "Community participation in a rural community health trust: The case of Lawrence, New Zealand", *Health Promotion International* 18(3): 189–197.

Farmer, J. and Kilpatrick, S. 2009. "Are rural health professionals also social entrepreneurs?", *Social Science & Medicine* 69(11): 1651–1658. doi: 10.1016/j.socscimed.2009.09.003.

Flannery, M. 2013. "Central Otago Health Incorporated". In Webb, A. (ed.) *Dunstan hospital clyde* (pp. 71–77). Cromwell: Central Otago Health Inc.

Flora, C.B. and Flora, J.L. 1993. "Entrepreneurial social infrastructure: A necessary ingredient", *The ANNALS of the American Academy of Political and Social Science* 529(1): 48–58. doi: 10.1177/0002716293529001005.

Haggerty, J., Campbell, H., and Morris, C. 2008. "Keeping the stress off the sheep? Agricultural intensification, neoliberalism and 'good' farming in New Zealand", *Geoforum* 40: 767–777.

Halseth, G. and Ryser, L. 2006. "Trends in service delivery: Examples from rural and small town Canada, 1998 to 2005", *Journal of Rural and Community Development* 1(2): 69–90.

Hanlon, N. and Halseth, G. 2005. "The greying of resource communities in northern British Columbia: Implications for health care delivery in already-underserviced communities", *Canadian Geographer* 49(1): 1–24.

Harvey, D. 2015. *Seventeen contradictions and the end of capitalism*. London: Profile.

Humphreys, J.S., Wakerman, J., Wells, R., Kuipers, P., Jones, J.A., and Entwistle, P. 2008. "'Beyond workforce': A systemic solution for health service provision in small rural and remote communities", *Medical Journal of Australia* 188(8): S77.

Johns, S., Kilpatrick, S., and Whelan, J. 2007. "Our health in our hands: Building effective community partnerships for rural health service provision", *Rural Society* 17(1): 50–65.

Kearns, A.A. and Joseph, A.E. 1997. "Restructuring health and rural communities in New Zealand", *Progress in Human Geography* 21(1): 18–32.

Kenny, A., Farmer, J., Dickson-Swift, V., and Hyett, N. 2015. "Community participation for rural health: A review of challenges", *Health Expectations* 18(6): 1906–1917.

KI (Key Informants). 2017. Interviews held with anonymous informants.

Larner, W. and Walters, W. 2000. "Privatization, governance and identity: The United Kingdom and New Zealand compared", *Policy and Politics* 28(3): 361–377.

Lynch, K. 2014. "'New managerialism' in education: The organizational form of neo-liberalism", *Open Democracy*. www.opendemocracy.net/kathleen-lynch/'new-managerialism. Accessed 16/3/2017.

MacLean, H. 2014. Community sees fruit in hard work. *Otago Daily Times* September 22. Available online at www.odt.co.nz/regions/south-otago/community-sees-fruits-its-hard-work. Accessed 19/8/2017.

MacLean, H. 2015. Home to be major employer. *Otago Daily Times* October 21. Available online at www.odt.co.nz/regions/home-be-major-employer. Accessed 19/8/2017.

Milligan, C. and Conradson, D. 2006. "The contemporary landscapes of welfare: The 'voluntary' turn". In Milligan, C. and Conradson, D. (eds.) *Landscapes of voluntarism: New spaces of health, welfare and governance* (pp. 1–14). Bristol, UK: Policy Press.

Mudge, S.L. 2008. "What is neo-liberalism?", *Socio-Economic Review* 6: 703–731.

Nel, E. 2012. *Responding to changing fortunes: The recent experience of small town New Zealand*. Paper presented at the New Zealand Geographical Society conference: Napier.

Nel, E. 2015. "Evolving regional and local economic development in New Zealand", *Local Economy* 30(1): 67–77.

North, P. 2002. "LETS in a cold climate", *Policy and Politics* 4(1): 483–499.

nzDoctor.co.nz. 27 April 2017. *Small West Otago town wins big at community awards*. Available online at m.nzdoctor.co.nz/.../small-west-otago-health-trust-wins-big-at-community-awards.aspx. Accessed 19/8/2018.

Peck, J. 2013. "Explaining (with) neoliberalism", *Territory, Politics, Governance* 1(2): 132–157.

Peet, R. 2012. "Comparative policy analysis: Neoliberalising New Zealand", *New Zealand Geographer* 68(3): 151–167.

Phillips, C. and McLeroy, K. 2004. "Health in rural America: Remembering the importance of place", *American Journal of Public Health* 94(10): 1661–1663.

Realpe, A. and Wallace, L. 2010. *What is co-production?* London, UK: Health Foundation. Available online at http://personcentredcare.health.org.uk/sites/default/files/resources/what_is_co-production.pdf.

Scoop.co.nz. 2015. *West Otago Health comes up as district's top volunteers*. Available online at www.scoop.co.nz/stories/AK1509/S00654/west-otago-health-comes-up-as-districts-top-volunteers.htm. Accessed 19/8/2017.

Seyfang, G. and Haxeltine, A. 2012. "Growing grassroots innovations: Exploring the role of community-based initiatives in governing sustainable energy transitions", *Environment and Planning C: Government and Policy* 30(3): 381–400.

Shone, M. and Memon, A. P. 2008. "Tourism, public policy and regional development: A turn from neo-liberalism to the new regionalism", *Local Economy* 23(4): 290–304.

Skinner, M.W., Joseph, A.E., Hanlon, N., Halseth, G., and Ryser, L. 2014. "Growing old in resource communities: Exploring the links among voluntarism, aging and community development", *The Canadian Geographer* 58(4): 418–428.

StatsNZ (Statistics NZ). 2018. *QuickStats about a place*. Available online at http://archive. stats.govt.nz/Census/2013-census/profile-and-summary-reports/quickstats-about-a-place. aspx?request_value=14999&parent_id=14990&tabname=#14999. Accessed 1/2/2018.

Strasser, R. 2003. "Rural health around the world: Challenges and solutions", *Family Practice* 20(4): 457–463.

Stuff.co.nz. 19 August 2017. *West Otago Health Trust celebrates success*. Available online at https://i.stuff.co.nz/southland-times/news/78477445/West-Otago-Health-Trust-cele brates-success-with-Tapanui-community. Accessed 19/8/2017.

Sullivan, L., Ryser, L., and Halseth, G. 2014. "Recognizing change, recognizing rural: The new rural economy and towards a new model of rural service", *Journal of Rural and Community Development* 9(4): 219–245.

Visser, R. 2013. "150 years of Dunstan hospital". In Webb, A. (ed.) *Dunstan hospital clyde* (pp. 117–118). Cromwell: Central Otago Health Inc.

Webb, A. (ed.) 2013. *Dunstan Hospital Clyde*. Cromwell: Central Otago Health Inc.

White, S. 2016a. Staff ready home for first residents. *Otago Daily Times* February 6. Available online at www.odt.co.nz/regions/south-otago/staff-ready-home-first-residents. Accessed 19/8/2017.

White, S. 2016b. Community inspiration celebrated. *Otago Daily Times* April 11. Available online at www.odt.co.nz/regions/south-otago/community-inspiration-celebrated. Accessed 19/8/2017.

Wilson, O.J. 1995. "Rural restructuring and agriculture-rural economy linkages", *Journal of Rural Studies* 11(4): 417–431.

6 Partnering for health care sustainability in smaller urban centres

Why and how a health authority chose to 'go upstream'

Neil Hanlon, Martha MacLeod, Trish Reay, and David Snadden

Introduction

Rural and remote small town settings are often difficult places in which to deliver health care because of their smaller populations and geographical remoteness. These challenges often contribute to, and are exacerbated by, poorer health outcomes among populations in these settings. At the same time, efforts to address rural health care delivery challenges too often focus on short-term patchwork fixes that, at best, restore and reproduce inadequate models of health care delivery.

Numerous commissions and task forces have called for more comprehensive strategies to tackle the multifaceted challenges facing rural health care (e.g., Health Council of Canada, 2005). Most notably, the Royal Commission on the Future of Health Care in Canada (more commonly known as the Romanow Commission) singled out the challenges of rural health care access across the country as needing a coordinated, national effort to find long-term solutions (Romanow, 2002). In the years following the Romanow report, most provinces and territories established working groups, developed policies, and published white papers and strategic frameworks intended to tackle the persistent problem of rural accessibility. These consistently called for a primary health care approach to rural and remote health care delivery, and for reforms such as team-based primary care practice, alternative models of physician remuneration, and more integrated and coordinated health care delivery (Health Council of Canada, 2005).

In spite of this consensus, progress in achieving these ideals in practice has, at best, been modest (Health Council of Canada, 2013; Hutchinson, 2008; Hutchinson *et al.*, 2011). Analysts point to many factors that create barriers to primary health care reform in Canada, but prominent among these are a lack of clear central policy direction, inadequate financial and administrative support, and strong resistance from health professional associations (Hutchinson *et al.*, 2001; Levesque *et al.*, 2015; Naylor, 1999). A reform initiative underway in northern BC, Canada, however, suggests that these barriers can be overcome. What began as an exercise in implementing pilot projects to improve chronic disease management became a collaborative, community-based, multi-sector, and interdisciplinary effort to achieve 'whole systems change' (Kodner, 2006). We are interested here in how and why a regional health authority felt it was important to 'go upstream' and engage with community leaders and service organizations as partners in this whole systems change initiative.

In particular, we present a case study of Northern Health's efforts to achieve sustainable, integrated, community-centred health systems, and discuss the challenges and benefits these efforts present for the health care sector and its community partners. Going upstream, in this case, entailed the creation of partnerships between the regional health authority, independent health care professionals (primarily physicians), and community leaders to pursue healthy community activities in conjunction with, and as part of, change management processes aimed at realigning community-based primary health care delivery. We thus explore the idea that engaging community leaders as agents of population health and primary health care transformation offers a way to address some of the more common challenges of health care delivery, especially in rural and remote settings.

Policy context

Health care policy in Canada bears the heavy imprint of a federalist political structure that tends to impede structural change (Davidson, 2004; Lazar, 2013). The *Canada Health Act* (Government of Canada, 1984) is arguably the most recent piece of federal health legislation of any major and enduring impact. The act amalgamated the *Hospital Insurance and Diagnostic Services Act* (Government of Canada, 1957) and *Medical Care Act* (Government of Canada, 1966), and affirmed the conditions by which provincial and territorial governments must abide in order to receive federal funding to help operate their respective single-payer health insurance plans. At about the time the *Canada Health Act* was passed, the federal government began steadily reducing its level of direct fiscal transfer from approximately 50% to just over 10% (Provincial and Territorial Ministers of Health, 2000); although, this figures does not take into account the value of federal tax points transferred to provincial governments over this time period (Lazar and Church, 2013). Provincial and territorial governments have since balked at any efforts of the federal government to expand the scope of Canadian medicare or alter its institutional structures (i.e., to reform its curative and acute-care orientation) (Lazar, 2013).

Provincial governments instead have been preoccupied with containing the rising costs of health care, since more of the responsibility to pay for these programs now rests with them. Provincial health reform efforts in these directions, especially in the areas of acute care rationing and rationalization, have met with considerable opposition from key interest groups (e.g., medical associations and nursing unions), as well as suspicion from the voting public that provincial reforms will erode the foundations of Canadian medicare. As a result, provincial governments regularly made bold policy pronouncements about wanting to 'restructure' and 'modernize' health care, including the kinds of comprehensive changes deemed critical to addressing rural health care challenges, but made little practical headway in actually doing so (e.g., Davidson, 2004; Hacker, 2004; Hanlon, 2017; Lazar, 2013).

The establishment of regional health governance in nine of ten provinces in the 1990s was meant to be a means to realign health care delivery from its acute care bias to more patient-focused health care organization (Lewis and Kouri, 2004; Lomas, 1997). Regional health governance took different forms across Canada, but the basic idea was that planning and priority setting would be devolved from provincial ministries of health to regional boards to ensure greater responsiveness to population needs. At the same time, control of facilities would be removed from local voluntary boards and centralized at a district or regional level as a means to rationalize acute care supply, redirect resources to traditionally underfunded sectors (e.g., mental health, home and continuing care, and public health), and enable more coordinated and integrated health care delivery across these different sectors (Casebeer, 2004; Touati et al., 2007; Trottier et al., 1999). For the most part, however, regionalization in Canada has not lived up to expectations, principally because health authorities have no direct authority over many key front line health professionals (e.g., family physicians), and because regional boards have typically been given little discretion by provincial ministries of health to pursue their own priorities for, and approaches to, health care reform (Barker and Church, 2017).

Yet one health authority, serving a predominantly rural and remote population, has managed to pursue a sustained strategic policy effort to bring about communitybased primary health care reform in conjunction and cooperation with independent health care practitioners and private service providers operating outside their chain of command. Below, we offer an account of the health authority's strategic efforts to bring about these health care transformations, with particular consideration of its decision to partner with municipal officials as a means to make these change management processes acceptable to health care professionals and relevant to community-based partners.

Case study: Northern Health's community-based approach to reform

The Northern Health Authority (hereafter referred to as Northern Health) was one of five regional health boards established by the BC government in 2002. It is responsible for administering a wide variety of health programs, including acute care, mental health and addictions, home and continuing care, public health, disease management, and diagnostic services. It also regularly consults with community groups and key stakeholders to set priorities for program planning, and is responsible for ensuring that its programming complies with the directives and guidelines of the provincial ministry of health services. The authority serves a relatively small population of fewer than 300,000 persons spread over a very large geographic area (approximately 600,000 km^2) (refer to Figure 6.1). This area includes only one urban centre with more than 50,000 people (i.e., Prince George, population 75,000), and five centres of between 10,000 and 20,000 residents. The remaining populations reside in settlements of between a few

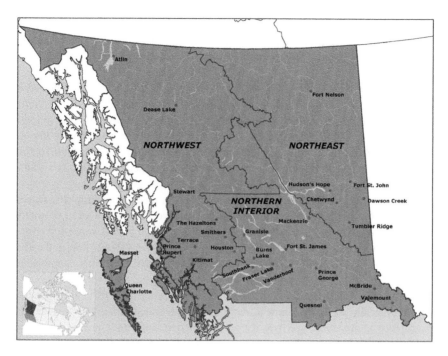

Figure 6.1 Map of the area served by Northern Health showing selected communities.
Map credit: Kyle Kusch

hundred and less than 10,000 persons. The health outcomes of the population in the region are generally among the poorest in the province, and many communities experience persistent problems recruiting and retaining health care personnel. In short, the population and communities served by Northern Health exhibit many of the classic challenges of rural and remote health care delivery.

The health reform initiative discussed in this case study was formalized in 2009, but its antecedents were evident years earlier. The case is therefore presented in three subsections: a preliminary and foundational phase (2002–2005), an adaptive strategy phase (2006–2009), and an experiential phase of partnering with community-based groups (2010 to present).

Foundations of community-based collaboration

Northern Health's formative years coincided with the completion of the Romanow Commission (2002) and the heightened interest that followed in 'renewing' Canadian medicare. While provincial governments balked at opting into national home care and pharmaceutical coverage plans, many did commit to pursue reforms that placed greater emphasis on primary health care principles. A major incentive for this was the Federal Government's Primary Health Care Transition Fund (PHCTF), which initially provided three years of funding to provincial governments to establish and study experimental pilot initiatives that promised to enhance access, improve quality, reduce inappropriate service utilization, and improve population health outcomes.

BC's share of the Fund was $74 million, of which $4.2 million was allocated to communities in the northern region. With these funds, the newly established Northern Health developed a three-year plan, *Improving our Health, Improving our System* (2003–06), that intended to focus exclusively on improved chronic disease management based on its own analysis of health outcomes data (Northern Health, 2006). Northern Health organized a five-day workshop in September 2003 to inform key stakeholders and 'community collaborators' (i.e., program managers, physicians, and health service administrators) of their plans for administering the federal pilot project funds. The feedback it received at these workshops, however, prompted an abrupt change of direction. That is, the community collaborators expressed strong concerns about the disease management focus of the Northern Health plan and called strongly for more community-based and population health oriented approaches (Northern Health, 2006).

Following this experience, Northern Health committed to a different approach to primary health care development that drew on integrated professional education, shared care approaches, the use of common patient information platforms, and public health education. The resulting Community Collaborative Project was designed around a model of community-based interdisciplinary provider teams that would manage wider areas of primary health care practice rather than chronic diseases exclusively. The newly redesigned pilot initiatives were then implemented in seven sites across the region. While these projects came to an end, their legacy is tied to the valuable lesson Northern Health learned about the importance of giving community stakeholders the opportunity to provide meaningful input about how to proceed with health care transformation.

The other key lesson from these pilot projects was about how a health authority could work collaboratively with physicians, who may otherwise be prone to regard regional health authorities as an unnecessary layer of bureaucracy or, worse, a threat to their professional autonomy and ability to act on behalf of patients (Davidson, 2004). Experiences in other provinces suggest that relations between physicians and health authorities had become more strained over time (e.g., Casebeer, 2004; Davidson, 2004; Reay and Hinings, 2009), but the newly created Northern Health at least had the advantage of a clean slate. The Community Collaborative initiative proved a useful opportunity for the newly formed health authority to leverage its resources to help support collective work already underway among the medical community. For instance, many physicians in the region's largest urban centre were using a common electronic medical record system. Northern Health used the pilot initiatives to support the continued diffusion and uptake of this system. Northern Health also committed to help support other efforts already underway in the medical community, such as practice support programs and locum coverage arrangements. These efforts paved the way for much greater collaboration and partnering in years to come.

While critics of pilot initiatives have decried their small scale and limited life span, the initiatives did create opportunities for collective learning and partnership building. This appears to be an important outcome of these projects in northern BC, as the following excerpt from Northern Health's (2006) final report of the PHCTF initiatives demonstrates:

> Northern Health and the medical community <u>learnt</u> [emphasis in the original] a number of essential 'lessons' as the result of the PHCTF initiatives, perhaps the most valuable being the importance of a positive working relationship between Northern Health and the family physicians providing primary health care across the health authority.
>
> (p. 73)

A second major lesson learned from the PHCTF was that communities should not be regarded simply as containers that receive health care in whatever form health administrators and health care providers deem to offer. Primary health care reform must include structures and processes for meaningful community engagement. As this excerpt from the same report states, "most [community collaborators participating in the workshop] felt they were not in a position to return to their communities to dictate or cajole massive system change" (p. 12). From this point on, Northern Health remained committed to engaging with community-based partners to ensure the legitimacy and sustainability of its health care transformation efforts.

An adaptive strategy

Even as additional federal funding was made available to pursue a second wave of primary health care pilots, Northern Health's executive was looking to plan for more sustained and comprehensive health care change. When its original PHCTF projects wrapped up in 2006, Northern Health began formalizing a strategy to guide reform efforts throughout the region. *Care North* was presented as a partnership between northern physicians and the health authority, guided by the vision of "[a] health care system founded in primary care and community, where every resident has a 'primary care home', providing access and comprehensive, coordinated care" (Northern Health, 2008, p. 10). Northern Health committed substantial internal funding to expand the number of communities and practices involved, expand its chronic disease management efforts, and establish an Integrated Health Network of physicians, nurses, pharmacists, and other health professionals to focus on so-called high-risk populations, such as frail elders and those living with mental health and addiction issues.

The need for community consultation and input was reiterated throughout the strategy, as illustrated in the following excerpt:

> If change is to occur, there will need to be considerable support for stakeholders at the local level to examine and evaluate different options/

opportunities, come to informed voluntary decisions as to the steps to be undertaken and to make (and evaluate) these changes.

(Northern Health, 2008, p. 31)

Nevertheless, 'communities' were not expressly named as co-partners in the *Care North* strategy. This may reflect a preoccupation at that time for Northern Health to ensure buy-in from physicians in the region. That is, national and provincial medical associations around this time had begun championing the concept of a medical home as a means to solve the problem of 'unattached patients' and create more sustainable conditions for general medical practice through better use of information technology, group medical practice, and enhanced practice support (College of Family Physicians, 2009, 2011). Concerned to keep the momentum of its reform efforts moving, Northern Health adopted the concept of a primary care home, which was clearly aligned to the vision of a medical home, and integrated this into its *Care North* strategy and vision for community-based, inter-professional primary health care teams.

In fact, other accommodations to physicians are evident in the *Care North* strategy, most notably in the interchangeable use of the terms primary care and primary health care. While this mixing of the terms is fairly commonplace in health care policy and literature, the terms differ in important respects (Muldoon *et al.*, 2006). Primary care, a term more generally favoured by the medical community, refers more narrowly to a level of care often envisioned as the 'point of first contact' between a patient and a general practitioner. Primary health care, on the other hand, is a more expansive term that refers to an approach to health service delivery and policy development emphasizing universal coverage, reducing health inequalities, *enhancing community participation in setting priorities* (emphasis added), and achieving inter-sectoral cooperation and commitment (Muldoon *et al.*, 2006).

In addition to its strategic accommodations to physicians in the region, Northern Health also proved adept at capitalizing on the momentum of macro-level cooperation between the provincial government and the BC Medical Association. In particular, the British Columbia Ministry of Health Service's (2007) *Primary Health Care Charter* set out broad strokes for greater collaboration between family physicians, health authorities, and key client groups (i.e., those living with, or at heightened risk of, chronic disease). The charter provided specific types of support in the form of investments in information management and information technologies, and dedicated funding for change management at the clinic and community levels through the establishment of regional practice support teams, which were to be administered by the regional health authorities. Northern Health deftly built this mandated responsibility to implement the Ministry of Health's practice support program into its *Care North* strategy. This proved valuable to the health authority's efforts to demonstrate its understanding of the pressures physicians faced, its commitment to offer support to physicians in private practice, and its ability to build trust with its physician partners at this critical early juncture in its primary health care strategizing.

By 2009, Northern Health had a completed final draft of its vision and strategy for achieving community-based primary health care transformation, centred around the concept of a primary care home. It then embarked on region-wide community consultations to share the draft strategy and solicit input and feedback from community stakeholders (Northern Health, 2009b). Following this, *Care North* was formalized and incorporated into the authority's new five-year strategic plan (Northern Health, 2009a).

Northern Health intended to carry out this work in three stages. The first stage was to engage with communities that had demonstrated the most willingness and preparedness for the initiative. Six communities were initially chosen for further primary health care development, and Northern Health set about recruiting 'change leaders' amongst local physicians and municipal officials (i.e., mayors or council members) in each of these six centres. From among its own personnel, Northern Health committed personnel in key local decision-making capacities to participate in implementation committees, providing background preparation in its goals and vision for reform. Each of these implementation committees was to meet regularly and begin the work of re-aligning local health services along the lines of a primary care home focus, but also to identify healthy community initiatives that would engage the entire community in wellness and health promotion.

The second stage of the process was to equip the implementation committees with information to understand the health needs and priorities of their respective communities in order to provide a clearer picture of existing service processes and capacities, and to enable the committees to begin considering how local care provision should be re-organized around the concept of primary care homes and inter-professional teams. Having municipal leaders as partners in these change management processes was a unique feature of the strategy, as well as an opportunity to include valuable 'community' insights and values in the proceedings.

The third stage was to be one of implementation committees identifying specific ways in which local services and resources were to be transformed to better meet the needs of the local community. This third stage was to have a health care change management component: that is, reorganizing and realigning health services delivered by Northern Health, and then integrating these with the services of physicians and independent providers organized specifically around the concept of the primary care home. The other key aspect of implementation, however, was that all local partners would work in partnership to design and deliver healthy community initiatives intended to have lasting effects on health promotion and health education in the community.

While Northern Health has subsequently had to adapt its plans, especially to accommodate physicians, it has taken care to preserve a collaborative and community-based focus to its approach. Community-based partners were not included simply to be kept abreast of health care changes; they were expected to help to determine the process by which these new care arrangements and alignments emerged. A clear indication of this community focus was that upstream healthy community initiatives were built into the health care change management process, and these were to be designed and delivered with the full participation of local physicians and health authority personnel rather than as strictly 'municipal' activities. This way of engaging with community is arguably the most distinctive feature of the Northern Health strategy, and most clearly expresses the health authority's commitment to community-based primary health care reform.

Upstream engagements

Health care boards and municipal officials in British Columbia do not have an extensive history of formal collaboration, except in the area of capital projects. The Province of BC's (1996) *Hospital District Act* mandates that health authorities and regional districts work together on the development of new or expanded hospitals and residential care facilities. Much of the extent of Northern Health's early experiences working with municipal leaders was thus tied to its work with regional hospital district committees comprised of elected officials from corresponding municipal governments. These hospital district committees are responsible for raising funds, acquiring land, and securing necessary land use planning approvals to enable Northern Health's capital projects to proceed.

With the growing interest in pursuing healthy community initiatives (e.g., 'town on a diet' and 'walk with a doc'), however, the municipal and health care fields are discovering wider areas of common interest. This is perhaps best exemplified by the World Health Organization's 'healthy cities' and 'healthy communities' movements of the past two decades in which decision makers from various fields and sectors work together to make communities more conducive to healthy living (Norris and Pittman, 2000). There has been a flourishing of Canadian healthy community initiatives in recent years, and some cities have gone so far as to adopt population health principles in their community plans and municipal service operations (e.g., City of Prince George, 2010).

Building on its experiences working with hospital district committees, Northern Health chose to recruit municipal officials, especially those with whom they had a history of positive working relations, to be the voice of communities in its primary health care transformation strategy. Northern Health created the Healthy Community Partnership (HCP) initiative as a commitment between a community leader (i.e., mayor or elected council member) and a Northern Health representative (e.g., often a health services administrator whose traditional purview is to oversee the management of health care programs and facilities) to work together to address health issues in the community. Once a municipal leader and health authority official agreed to form a partnership, they worked to establish a healthy community committee, recruit representatives, study local health issues, identify vulnerable populations, and develop and implement an action strategy to address whatever upstream causative factors had been targeted. The mayor or elected councilor also committed to sit on the local implementation committee to help ensure the success of upstream projects, but also to participate actively in the health care change management processes underway.

The decision to recruit municipal leaders as participants, rather than members of the public or representatives of health care client groups, was meant to enhance the legitimacy and visibility of the Healthy Community Partnership. From a practical standpoint, this was also the surest way to ensure that the implementation committees had access to their respective municipal government resources and personnel. While it could be argued that Northern Health's approach is yet another instance of health care reform without the direct input or involvement of the public, there is also a case to be made that elected officials are supposed to be well informed of the public interest, and are obligated to act in the interests of their electorate.

In any case, Northern Health's healthy communities initiative flourished. According to an internal report, healthy community committees had been established and were functioning within the first year in all six communities chosen by Northern Health to lead the reform initiative. Mayors of several other communities in the region expressed interest in creating Health Community Partnerships, even though their communities were not among these original six reform sites. By the following year (i.e., 2012), the number of partnerships reported to be in various stages of establishment had grown to eleven. All eleven committees had held initial meetings, nine had identified the community's risk factors, eight had adopted terms of reference or partnership agreements, eight had applied for community grant funding from Northern Health, and one had submitted a strategic plan.

This apparent success of Northern Health's partnership with communities may have been due, in part, to the fact that a parallel program was being offered simultaneously through the British Columbia *Healthy Communities* initiative funded by the BC Ministry of Health (www.bchealthycommunities.ca). The province-wide program was promoted as BC's participation in the international 'healthy communities' movement, and involved extensive collaboration between the BC Ministry of Health, the Union of BC Municipalities, health authorities, and various community groups. At the time the Northern Health program was being implemented, there were three northern communities participating in the province-wide initiative. This may have led to some confusion and conflation of the two programs, but it does not detract from the high level of interest and commitment from municipal leaders throughout the region, nor from the opportunities that Northern Health's program created for greater collaboration, information exchange, networking, and co-learning with the municipal government and community service sectors.

The engagement of municipal leaders in 'healthy communities' initiatives also appears to have awakened, at least in some municipal leaders, a greater sense of agency in shaping the future of health care in their communities. In one of the six communities chosen to lead the reform initiative, municipal leaders were mobilized to help deal with a crisis all too common in the region: the sudden departure of a number of physicians in the community. In this instance, municipal leaders, in conjunction with health authority partners, ministry officials, and local physicians, worked collaboratively to establish a non-profit primary care society, negotiate alternative physician funding arrangements, recruit new physicians, and open a new multidisciplinary clinic intended to serve as a primary care/medical home for the community.

Discussion

All of the familiar policy challenges described earlier (i.e., lack of clear and consistent policy direction from senior levels of government, and potential resistance from health professionals) were present to varying degrees in this case. These structural impediments may help explain why a health care transformation initiative that began in 2011 is still considered to be in its implementation phase. Whatever else comes of these efforts, however, municipal leaders are engaged in, and remained committed to, the process and appear to have embraced their new role in primary health care development.

The case also highlights the importance and value of persistence. Northern Health remained committed to engaging local stakeholders both within and outside the health care field as a means to ensure that reforms remain consistent with a community-based population health approach. It took its own initiative to move reform beyond the stage of pilot project experimentation and towards 'whole systems' change and has remained committed to its vision for community-based primary health care reform (e.g., Northern Health, 2016). Finally, as we have attempted to articulate in this study, the health authority exhibited an ongoing effort to ensure that communities were made full partners in these reform efforts.

A series of factors helped sustain the initiative. First, Northern Health consistently altered its plans and approaches to align with macro-level events (e.g., new provincial policy) and capitalize on opportunities to build and strengthen its partnerships with physicians and communities. Second, Northern Health has enjoyed an uncommonly lengthy period of stable leadership (i.e., more than a decade) during which time the senior executive team has enjoyed the full confidence of its board of governors. Third, the health authority structure itself has not been altered or restructured in any major way, which is also rare among regionalized health systems in Canada (Barker and Church, 2017). While this combination of factors is not likely to be present elsewhere, the case as a whole does offer compelling support for the notion that health care reform requires time, persistence, and opportunism.

This is not to say that Northern Health's reform approach is without flaws and blemishes. There were missteps and backtracks along the way, and progress has been considerably slower than originally envisioned. Northern Health has encountered a good deal of internal resistance, for instance from unions, which has slowed its own efforts to realign local services. As a result, the time commitments of community partners have been greater than first anticipated, and this comes at a cost to all concerned. We have suggested that municipal leaders, physicians, and other community-based groups are generally on board with Northern Health's plans and approach, but this is certainly not universally the case. We have intentionally omitted accounts of resistance and pushback in order to focus on the unique community-oriented aspects of Northern Health's reform strategy.

These shortcomings and setbacks aside, the work of partnering for community-based primary health care reform in northern British Columbia continues. In its most recent five-year strategic plan, Northern Health reiterated that the commitment to establish primary care homes and inter-professional teams in urban centres throughout the region remained its highest strategic priority (Northern Health, 2016). Of special note, for our purposes, is that its efforts to bring municipal leaders from across the region on board as active partners in health care reform continue to make headway. As of April 2018, Northern Health reports it has actively engaged 20 communities in Healthy Community Partnerships (Northern Health, 2018). This figure represents approximately two-thirds of all settlements of more than 1,000 persons in the region. Thus, while Northern Health's ultimate objective to achieve 'whole systems change' remains a work in progress, there is evidence that its 'upstream' connections are gaining momentum, which should bode well for the prospects of ongoing health care reform.

Conclusion

In this final section, we consider lessons that this case offers for those interested in pursuing similar types of health care reform in other remote and smaller urban jurisdictions. The first is the importance of approaches that respect and reflect the conditions, needs, and preferences of the places affected. Communities cannot be treated as mere backdrops to health care reorganization and reform. A major part of this is to make sure that members of local communities have reason to feel they have a sense of ownership over their health care.

The second lesson is that health care reform must look beyond physicians and health care administrators as agents of change. In this case, community leaders were recruited as partners in change management, but this is not to suggest this is the best or only means to invite participation from community members. The case highlights the considerable and sustained efforts on the part of health administrators to ensure there were adequate opportunities and platforms for community engagement and joint planning. For their part, municipal leaders stayed committed to the process over an extended period of time and stepped up at critical junctures to keep things moving.

A third lesson is that community-based reform efforts are not sufficient on their own. A certain level of central oversight is necessary, be that from a regional body or higher. On a more practical level, local agents of change have to be able to count on informational and financial resources that are beyond their means. Having strong central policy and support for health care reform is a major challenge in Canada for reasons described earlier in the chapter (i.e., federal-provincial stalemate, resistance from vested interests). In this case, the stability of the health authority and its persistence in pushing a particular vision for primary health care reform, in collaboration with its communities, proved sufficient to build and maintain a certain level of momentum and support from central policy.

Fourth, this case speaks to the importance of adaptive strategy and collective learning. All partners in the process made accommodations and learned from their experiences. The health authority learned important lessons about community consultation and how to work with those outside its organizational domain. Physicians learned to relate differently to communities and to other health professionals and service providers. Finally, municipal leaders appeared to enhance their awareness of the intricacies of the health care field, and grow more confident in their role as partners in reform.

The case of primary health care reform in northern BC is by no means an obvious success. The strengths we highlight are, to date, more structural and procedural than outcome-based. The main achievements to date are that physician groups are engaged, a space has been created for local community leaders to have a meaningful say in health care governance, and some headway has been made in making change in rural health care delivery. Moving forward, these efforts will need a balance of central direction and local discretion to ensure that health care is realigned and re-oriented to best reflect the needs, preferences, and circumstances of the communities served.

References

Barker, P. and Church, J. 2017. "Revisiting health regionalization in Canada: More bark than bite?", *International Journal of Health Services* 47(2): 333–351.

British Columbia Ministry of Health Services. 2007. *Primary health care charter: A collaborative approach.* Victoria, BC: BC Ministry of Health Services. Available online at www.health.gov.bc.ca/library/publications/year/2007/phc_charter.pdf, 3 April 2018.

Casebeer, A. 2004. "Regionalizing Canadian healthcare: The good – The bad – The ugly?", *Healthcare Papers* 5(1): 88–93.

City of Prince George. 2010. *myPG social development strategy.* Prince George: The City of Prince George. Available online at: www.princegeorge.ca, 6 April 2018.

College of Family Physicians. 2009. *Patient-centred primary care in Canada: Bring it on home.* Mississauga, ON: College of Family Physicians.

College of Family Physicians. 2011. *A vision for Canada: Family practice: The patient's medical home.* Mississauga, ON: College of Family Physicians.

Davidson, A. 2004. "Dynamics without change: Continuity of Canadian health policy", *Canadian Public Administration* 47(3): 251–279.

Government of Canada. 1957. *Hospital Insurance and Diagnostic Services Act.* Statutes of Canada, 5–6 Elizabeth II (c. 28, s 1). Ottawa: Government of Canada.

Government of Canada. 1966. *Medical Care Act.* Statutes of Canada (c. 64, s 1). Ottawa: Government of Canada.

Government of Canada. 1984. *Canada Health Act, Bill C-3.* Statutes of Canada, 32–33 Elizabeth II (RSC 1985, c. 6; RSC 1989, c. C-6). Ottawa: Government of Canada.

Hacker, J.S. 2004. "Dismantling the health care state? Political institutions, public policies and the comparative politics of health reform", *British Journal of Political Science* 34(4): 693–724.

Hanlon, N.T. 2017. "Putting preservation first: Assessing the legacy of the Campbell government's approach to health policy". In J.R. Lacharite and T. Summerville (eds.), *The Campbell revolution? Power, politics, and policy in British Columbia* (pp. 150–164). Montreal: McGill-Queen's University Press.

Health Council of Canada. 2005. *Primary health care – A background paper to accompany health care renewal in Canada: Accelerating change.* Toronto: Health Council. Available online at www.healthcouncilcanada.ca.

Health Council of Canada. 2013. *Progress report 2013: Health care renewal in Canada.* Toronto: Health Council. Available online at www.healthcouncilcanada.ca.

Hutchinson, B. 2008. "A long time coming: Primary healthcare renewal in Canada", *Healthcare Papers* 8(2): 10–24.

Hutchinson, B., Abelson, J., and Lavois, J. 2001. "Primary care in Canada: So much innovation, so little change", *Health Affairs* 20(3): 116–131.

Hutchinson, B., Levesque, J.F., Strumpf, E., and Coyle, N. 2011. "Primary healthcare in Canada: Systems in motion", *The Milbank Quarterly* 89(2): 256–288.

Kodner, D.L. 2006. "Whole-system approaches to health and social care partnerships for the frail elderly: An exploration of North American models and lessons", *Health and Social Care in the Community* 14(5): 384–390.

Lazar, H. 2013. "Why is it so hard to reform health-care policy in Canada?" In H. Lazar, J.N. Lavis, P.G. Forest, and J. Church (eds.), *Paradigm freeze: Why it is so hard to reform health-care policy in Canada* (pp. 1–20). Montreal: McGill-Queen's University Press.

Lazar, H. and Church, J. 2013. "Patterns in the factors that explain health-care policy reform". In H. Lazar, J.N. Lavis, P.G. Forest, and J. Church (eds.), *Paradigm freeze: Why it is so hard to reform health-care policy in Canada* (pp. 219–252). Montreal: McGill-Queen's University Press.

Levesque, J.F., Haggerty, J.L., Hogg, W., Burge, F., Wong, S.T., Katz, A., Grimard, D., Weenick, J.W., and Pineault, R. 2015. "Barriers and facilitators for primary care reform in Canada: Results from a deliberative synthesis across five provinces", *Healthcare Policy* 11(2): 44–57.

Lewis, S. and Kouri, D. 2004. "Regionalization: Making sense of the Canadian experience", *Healthcare Papers* 5(1): 12–31.

Lomas, J. 1997. "Devolving authority for health care in Canada's provinces: 4. Emerging issues and prospects", *Canadian Medical Association Journal* 156: 817–823.

Muldoon, L.K., Hogg, W.E., and Levitt, M. 2006. "Commentary: Primary care (PC) and primary health care (PHC): What is the difference?", *Canadian Journal of Public Health* 97(5): 409–411.

Naylor, C.D. 1999. "Health care in Canada: Incrementalism under fiscal duress", *Health Affairs* 18(3): 9–26.

Norris, T. and Pittman, M. 2000. "The healthy communities movement and the coalition for healthier cities and communities", *Public Health Reports* 115: 118–124.

Northern Health. 2006. *Northern Health review of PHCTF funded initiatives*. Prince George, BC: Northern Health.

Northern Health. 2008. *Care North: Our health. Our future. Working together for accessible, comprehensive and effective primary health care services*. Prince George, BC: Northern Health Authority.

Northern Health. 2009a. *Let's talk about primary health care: Report on community consultations 2009*. Prince George, BC, Canada: Northern Health Authority.

Northern Health. 2009b. *Strategic plan – 2009 to 2015*. Prince George, BC: Northern Health Authority.

Northern Health. 2016. *Strategic plan … looking to 2021*. Prince George, BC, Canada: Northern Health Authority.

Northern Health. 2018. *Northern Health overview*. Prince George, BC, Canada: Northern Health Authority.

Province of British Columbia. 1996. *Hospital District Act*. Victoria, BC: Queen's printer. Available online at www.bclaws.ca/civix/document/id/lc/statreg/96202_01, 6 April 2018.

Provincial and Territorial Ministers of Health. 2000. *Understanding Canada's health care costs: Final report*. Toronto: Ministry of Health and Long-Term Care.

Reay, T. and Hinings, C.R. 2009. "Managing the rivalry of competing institutional logics", *Organizational Studies* 30(6): 629–652.

Romanow, R. 2002. *Building on values. The future of health care in Canada – Final report.* Ottawa: Commission on the Future of Health Care in Canada.

Touati, N., Roberge, D., Denis, J.L., Pineault, R., and Cazale, L. 2007. "Governance, health policy implementation, and the added value of regionalization", *Healthcare Papers* 2(3): 97–114.

Trottier, L.H., Champagne, F., Contandriopolous, A.P., and Denis, J.L. 1999. "Contrasting visions of decentralization". In D. Drache, D. Sullivan, and T. Sullivan (eds.), *Health reform: Public success, private failure* (pp. 147–165). London: Routledge.

7 Residential models of dementia care in rural Australian communities

Rachel Winterton, Kaye Knight, Catherine Morley, and Wendy Walters

Introduction

A key global challenge for rural health and residential aged care services in the future will be their ability to cater for the needs of people with dementia. Globally, approximately 47 million people are living with dementia (Alzheimer's Disease International, 2015). Given that older adults are proportionally more likely to live in rural and regional settings (Glasgow and Brown, 2012; Skinner and Hanlon, 2016), the number of older people living with dementia in rural settings is likely to increase (Stewart *et al.*, 2016).

However, social and environmental barriers related to rural living pose significant challenges for older adults living with dementia in rural settings, in terms of accessing services to enable them to age in place. Rural-community-dwelling older adults living with dementia face challenges accessing and using appropriate dementia care services. Stigma attached to dementia and lack of privacy within local services can be disincentives to seeking support. This situation is compounded by limited access to specialists in dementia assessment, management services, and local respite and day programs, and greater costs associated with travel to access non-local services and respite. This places additional burden on families and reduces the capacity and desire of older people living with dementia to access services, which often leads to a crisis that necessitates admission to a residential aged care facility (Dal Bello-Haas *et al.*, 2014; Herron and Rosenberg, 2017). Consequently, as Morgan *et al.* (2015) have noted, rural communities, and residential aged care facilities (RACFs) by extension, must be prepared to care for increasing numbers of people with dementia.

Since the early 1990s, management of dementia within long-term care has moved from an increasingly medicalized model toward a more social, person-centred model of care, which emphasizes the continuation of self and identity while seeking to reduce overprescribing (Edvardsson *et al.*, 2010; Saleh *et al.*, 2017). A 2015 Australian report has noted that 52% of older adults in permanent RACFs have a dementia diagnosis and require a higher level of care than people without a dementia diagnosis (Australian Institute of Health and Welfare, 2016). As Catic *et al.* (2014) note, RACF residents with a dementia diagnosis frequently demonstrate behaviour problems, which pose additional challenges for rural RACFs. Rural RACFs face significant resource and operational challenges as a result of increased workforce costs associated with geographical remoteness, limited catchment areas and lack of critical mass, limited internet coverage, increased travel, and freight costs (Australian Government, 2016). International research suggests that rural RACFs are subsequently less likely to have specialist dementia care units and may have knowledge deficits around caring for people living with dementia (Henderson *et al.*, 2016; Morgan *et al.*, 2015). Further, while challenging behaviours associated with dementia impact on job stress, burnout, and employee turnover within RACFs (Morgan *et al.*, 2005; Talbot and Brewer, 2016), this is problematic in rural contexts due to existing issues with recruiting and retaining staff to work in rural RACFs (Warburton *et al.*, 2014). Therefore, there is a need to explore how person-centred care for older people living with dementia can be delivered within both rural and urban RACFs, and this chapter details the implementation of an innovative person-centred model of care within one small rural Australian community.

The ABLE-D model

The ABLE model was designed and implemented in 2012, and further developed into ABLE-D in 2017, by a small rural health service, Rural Northwest Health (RNH). RNH is a public-sector health and aged care service located in western Victoria, Australia's southernmost mainland state. In Australia, rural aged care services are primarily operated by the public and community sector, with 91% of providers in outer regional, remote, and very remote locations operated by government and not-for-profit organizations (Baldwin *et al.*, 2013).

RNH provides a range of acute, aged, and primary health care services within the local government area (LGA) of Yarriambiack, which is located approximately 350km from the state capital city of Melbourne (see Figure 7.1). The Yarriambiack Shire has a population of 6,738 and covers a large geographical area of 732,407 hectares, with a comparatively low population density in relation to the rest of Victoria (0.01 persons per hectare). It is primarily comprised of rural land, with one large residential centre (Warracknabeal – 2,400 persons) and a series of smaller townships of less than 1,000 persons. The primary industry of employment is agriculture, with land used largely for sheep and grain growing (wheat and barley). As with many other Australian agricultural regions, the population has declined significantly over time, with significant population loss from the early 1990s. The population is also ageing more rapidly than its rural counterparts, with 19.5% of the population aged over 70 years, compared with 13.8% across regional Victoria (ProfileID, 2017). While Victoria is relatively urbanized in relation to other states of Australia, Yarriambiack is located in one of Victoria's most remote health care regions. The Modified Monash Model (MMM) is used to assess geographical remoteness in relation to health service provision in Australia, and it takes into account both geographical remoteness and population size. It has seven categories (MMM1–MMM7), with higher numbers denoting higher levels of rurality (Department of Health, 2017). Under the MMM model, the Yarriambiack

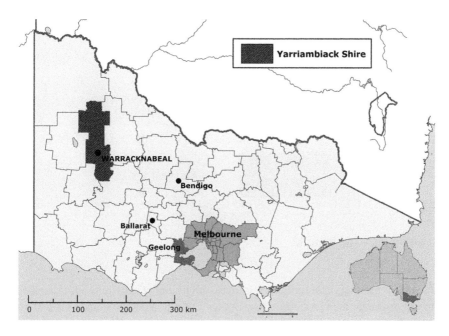

Figure 7.1 The local government area of Yarriambiack
Map credit: Kyle Kusch.

Shire is primarily classified as MMM5, with some of its outlying small townships classified as MMM6.

RNH has three campuses across the Yarriambiack Shire – one in the primary population centre (Warracknabeal) and two within smaller, more remote townships (Beulah and Hopetoun). Their RACFs are located at Warracknabeal (60 beds) and Hopetoun (24 beds), with a 15-bed secure dementia facility at Warracknabeal referred to as the memory support unit (MSU). It is within this unit that ABLE-D was implemented.

The ABLE-D model was designed with the aim of improving dementia care to residents within the MSU by means of extending and retaining residents' existing capabilities and maximizing quality of life. Team members within the MSU had observed a number of problematic behaviours among residents, which suggested that their needs were not being adequately met. These behaviours included pacing, wandering, physically and verbally aggressive behaviour, observed boredom, and high levels of daytime sleeping. ABLE-D's design was based on a person-centred approach that incorporates Montessori principles, which are gaining increasing traction within dementia care practice (Sheppard *et al.*, 2016). The model sought to facilitate a number of system changes at an organizational level and, as detailed in Figure 7.2 and Table 7.1, the model encompasses five core areas of focus: abilities and capabilities; background of the resident; leadership,

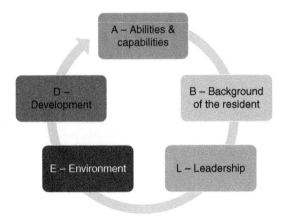

Figure 7.2 The ABLE-D model

education, training, and cultural change; environment that provides cues to support resident memory; and development of the condition and our understanding. While the emphasis on capability within dementia care is not new (Moyle *et al.*, 2016), the innovation of ABLE-D lies in its 'whole of model' approach, its promotion of a long-term view of cultural change at every level of the organization, and a strong focus on giving residents purpose. The model facilitates a feedback loop as the needs of residents develop in new ways, and this new data informs each focus area. This feedback loop is important in the rural context. Given the lack of specialist dementia and geriatric services in rural and remote Australia (Hansen *et al.*, 2005), staff are often not aware of the underlying disease that causes the dementia in all individuals. By including the component of development, the whole team is open to discovering new ways of thinking, doing, and learning to make life better for residents.

The implementation of this model has had numerous benefits for older adults within the MSU, which have included

- Cessation of prescribed psychotropic medications and significant reduction in prescription of other sedatives,
- Reduced physically non-aggressive and verbally agitated behaviour,
- Increased person-centred care practices by team members,
- Increased engagement of residents in meaningful activities, and
- Positive perceptions of family members regarding the homelike nature of the unit, and their trust in the care their family members were receiving (Roberts *et al.*, 2015).

The application of this model has been sustained since its implementation in 2012, with the model currently being adapted for use within community care

Table 7.1 Components of the ABLE-D model (adapted from Roberts *et al.*, 2015).

Abilities and capabilities	• Assessment of cognitive and physical capabilities using standard clinical assessment processes which are used to encourage independence and meaningful activity in areas of capability.
Background of the resident	• Collection of background information about residents (e.g., life stories and interests, likes, dislikes, skills, and dreams) via discussions with residents and their families, as well as written and visual displays.
	• Used to enable team members to support residents to undertake appropriate and meaningful roles and activities within the scope of their capabilities.
Leadership	• Leadership from the board, chief executive officer, managers, and non-managerial leaders within the care team to facilitate changes in process and practice.
	• Education sessions conducted by a dementia consultant for all nursing, care, and environmental services team members working in the MSU (two days of dementia care training and two days of Montessori activity training).
	• Ongoing support from the dementia consultant.
	• Support from the Cognitive Rehabilitation Therapist (CRT) and project manager provided to other team members.
Environment	• Transition to a more homelike environment in order to support the memory of the residents and enhance their capacity.
	• Installation of new signage, large-print name badges for team members.
	• 'Interactive' wall space introduced.
	• Adaptation of existing internal spaces to fit the needs of residents (addition of a small shop and a relaxation room for residents; removal of the television from the common area; and establishment of specific areas for music, hobbies, quiet reflection and reading, physical activity, games, storytelling, social interaction, and domestic activities).
	• Adaptation of the external environment to enable greater interaction and synergy with rural living environments (inclusion of a pergola, an old car, a chicken coop with chickens, a mural wall, gardening spaces, various private seating areas, and a barbeque).
Development of the condition	• Discovery of new things about a person as the disease progresses.
	• Developing understanding, ways of thinking, practice, and individual plans of care to continue to meet the changing needs of the residents.
	• Designing new theoretical thinking.

settings. Consequently, while this model has had significant successes in a clinical sense, the subsequent section explores some of the barriers and enablers associated with implementing and sustaining this model in rural contexts.

Factors associated with successful implementation and sustainability

Resourcing the program

As with any rural initiative, a key issue in developing new models of care relates to securing financial support (Kim *et al.*, 2017). In implementing the ABLE-D model, funding and staffing time were required to conduct refurbishment of the MSU, to reflect the new model of care, to implement the new model, and to train all clinical and non-clinical team members working in the MSU in Montessori for Dementia and Behaviour Management Education. However, a key success factor related to the ability of RNH to leverage existing resources and community partnerships to secure the required staffing and funding. Therefore, the support and leadership from the chief executive officer and board of management to support model implementation and the associated culture change were critical.

Public residential aged care services in Australia are funded by the federal government using the Aged Care Funding Instrument (ACFI), which measures care levels for each resident and focuses on care required to manage a person's disability rather than ability (Fine and Yeatman, 2009). However, hospitals and health services are funded through a mix of activity-based (funding attached to the delivery of specific services) and block funding. Block funding represents a system of funding hospital services based on a fixed amount, which is subject to certain conditions but not directly linked to either an amount or type of service provided. In Australia, small rural hospitals (such as RNH) are often funded through a block funding agreement, in recognition of the fact that activity-based forms of funding are not always feasible in areas of low critical service mass (Department of Health and Human Services, 2016; Duckett and Breadon, 2014). This funding flows from the federal government to state-level health departments, who manage and administer the funds directly to organizations, and allows greater flexibility in terms of funding allocations (Department of Health and Human Services, 2016; Fine and Yeatman, 2009). As a small rural health service providing both acute and aged care services, this funding model gave RNH the flexibility to divert some funds to finance a 0.5 FTE project manager to initiate the program, with this team member already being an existing member of the aged care team.

Funding for initial training was sourced through a grant obtained by Health Workforce Australia, a now-defunct federal government statutory authority focused on health workforce reform, which allowed for RNH team members to undertake the specialized Montessori training integral to the model. Team member responsibilities were also amended to meet the new needs of the model, within existing staffing models. For example, it was determined that cognitive rehabilitation therapist roles were needed, and so existing enrolled nurses employed within the MSU were rebranded and trained in relation to their changing role (to focus on memory support rather than behaviours). Once established, the model of care has been able to function within the allocated ACFI funding levels provided, with the day-to-day care associated with model implementation not requiring additional staffing. The ability of the program to be sustained without these additional resources is attributed to a cultural change within the organization, associated with strong leadership.

However, the fit of the ABLE-D model with regulatory health guidelines was a key challenge in resourcing some of the key objectives associated with implementation of the model. A key example related to regulations around infection control in aged care facilities, with some of the activities associated with the delivery of the ABLE-D model (e.g., the installation of a sensory wall and allowing residents to assist with meal preparation) posing challenges to meet infection control guidelines. This required dedicated work from the MSU team to challenge existing protocols and ensure that these guidelines could be met. For example, residents involved in meal preparation and serving were trained by team members to adopt organizational-standard hand hygiene practices, with hand hygiene rubs placed within easy access and accompanied with signage to guide use. Data were also collected and used successfully to demonstrate that there was no change in infection rates associated with incorporation of residents in meal preparation.

External collaborations

In the rural context, collaborations and networking are critical in accessing resources and expertise to implement sustainable local models of health care (Singh *et al.*, 2010). While this was also the case in implementing ABLE-D, these collaborations and networks ranged from local and regional to national. At the local level, collaborations and networking with local GPs were critical to successful implementation of the project. Initially, local GPs exhibited resistance to the ABLE-D model, largely due to resistance to change in processes around psychotropic prescribing and referrals to geriatricians. However, this was addressed through relationship building and education strategies, with the local GPs now supportive of the ABLE-D model processes. At the regional level, the initial grant application to Health Workforce Australia to access funding for the Montessori training involved two other small rural health services as partners. Other rural health care organizations have also visited RNH to learn about the model and to provide feedback on the model, and as a consequence they are currently developing and delivering their own version of the organization's model.

Given the innovation of the model and the lack of local expertise in Montessori, a great deal of support from state and national advocacy and professional organizations was required to access relevant training and expertise. Advocacy and professional organizations such as Montessori for Dementia and Dementia Australia were called upon to provide training, mentoring, and expertise. Dementia Australia, the national advocacy body for dementia, provided a trainer and mentoring support, conducted visits to RNH, and provided access to low-cost or free resources. Universities and consultants were also utilized at various stages of project design to ensure that the model was designed in a way that was clear and easily communicated, and to aid with knowledge translation within community contexts. This represented not only the ability to communicate how this model could benefit local community members and their families (in light of a change to a service that was familiar to, and well respected within the community), but also the ability to conduct knowledge translation to other rural hospitals and settings. The intention was that other communities could then adopt the model and amend it to be context-specific, while still keeping the core elements of the ABLE-D model.

Team member and community buy-in

In implementing innovative models of health or aged care, the readiness of staff to adopt and implement these models is critical to sustainability (Christi *et al.*, 2010). In the case of ABLE-D, to enable this model to work within a small integrated service providing acute, aged, and community care, team members from various departments across the organization (including nursing, food services, cleaning, and maintenance) were required to invest in the development and implementation of the model. This investment included modification of existing work practices and adherence to procedures associated with implementation of the ABLE-D model.

A key success factor in ensuring that team members were familiar with and supportive of the model was the provision of effective, timely training. As part of the initial implementation, all team members in the MSU, as well as the broader residential care facility, completed the full Montessori for dementia training, with the CEO and team members in the co-located acute hospital completing a one-day training course. All team members were also included in co-design activities associated with the development of the model, including brainstorming of refurbishment activities within the MSU. Leadership was also a key component of the successful implementation of the model, and the trust associated with this leadership, with some team members resistant to change and unwilling to try something new. Consequently, a dedicated and visionary project manager was essential to successful implementation, in addition to the presence of unwavering support from the CEO and board of management. As a small rural health service, the RNH board is largely comprised of community members, including local farmers, business owners, professionals, and relatives of people living within the MSU. These individuals had the primary responsibility for driving the model within the organization, which was essential in obtaining buy-in from team members. Getting the right people that wanted to work within the MSU was also critical in implementing the model successfully, with team members required to be invested in the model and willing to work in the MSU, and having the ability to think creatively in relation to how to meet patient needs in light of institutional rules and regulations. They also needed to be resilient, particularly with initial negativity from some team members external to the MSU in relation to perceived preferential resourcing. Over time, the outcomes associated with implementation of the ABLE-D model have also proved successful in maintaining team buy-in. The model has received numerous state health and aged care awards, which has attracted other health services to visit RNH and the MSU, and consequently instilled a sense of pride and purpose. Seeing positive outcomes for residents that have occurred as a consequence of the model has also been beneficial in articulating the difference the model has made.

On an ongoing basis, fundraising to improve the unit has also provided significant impetus to sustain community engagement in the model. While it is common in Australia for small rural health services to provide dementia care, in light of the high rates of people living with dementia in these regions, they tend to be based on comfort-based care rather than delivery of active memory support within specialized units. Therefore, this fundraising was critical not only in enabling the development of the model's objectives within the facility, but also in promoting community engagement and raising awareness. Fundraising efforts are largely driven by RNH, with support from the board of management (which, as previously stated, is comprised of local community members). However, these fundraising efforts have been well supported by the broader Yarriambiack community. A fundraising walk, "Walk for Yarriambiack", is held annually, with all proceeds diverted to the residential aged care unit. Team members within the MSU and wider residential aged care facility also run regular community raffles, with prizes donated by team members and local community organizations. As the physical facility improvements implemented through initial fundraising efforts became more prevalent (e.g., development of sensory gardens), family members of residents of the MSU also began to donate money to ensure continued development and upkeep of the model.

Numerous studies have noted the importance of community buy-in to the success of new rural models of health care (Vines *et al.*, 2016), and the continued engagement of the friends and relatives whose loved ones lived in the unit, in addition to the wider community, was essential for the success of the model. Families being inducted into the MSU receive considerable information on the ABLE-D approach prior to admission, and they receive continued contact with the cognitive rehabilitation specialist, which builds trust in relation to the model. Relatives of residents can attend the Montessori training, which also helps them to understand the principles guiding the care of their loved ones. The Montessori for Dementia consultant travels interstate up to four times a year to visit RNH and provide team member training. During her stay, she works with the MSU team to arrange the training plan and provide additional training of relatives and other community members as required. The cost of this training is free to relatives, and is prioritized within the annual RNH training budget. Providing this training is seen as highly beneficial in that it assists RNH to better communicate, and partner, with relatives. Additionally, the better informed they are about the care of their relative, the more meaningful their contributions to developing the service become, with relatives also included in all co-design activities pertaining to the MSU in conjunction with team members and residents. Within the wider community, RNH conducts regular dementia awareness education, which has enabled a greater community understanding of the principles underlying the model.

Key challenges associated with obtaining buy-in were primarily associated with two factors. First, successful development and communication of the model prior to implementation were critical, and challenges were associated with attempting to implement changes and provide education before defining and articulating the model to team members and families. Some of the initial training conducted prior to clear articulation of the model had to be repeated, as team members were unsure how to implement the training they had received without the appropriate framework and leadership. Other challenges included the movement of team members in and out of the MSU, which meant that regular training and reinforcement of the model were required, and multiple team members within the MSU were needed to take the lead in driving the model to ensure sustainability. This presented a challenge in regard to organizational capacity to grow leaders to drive the model. Additionally, the dual relationship of rural health and aged care service providers as both community members and health care professionals (Barrett *et al.*, 2016) was both an enabler and a barrier in regard to community support. It meant that while positive support for the model was easily translated into the community, feelings of negativity in relation to the model could also be easily communicated. Consequently, the MSU team needed to provide regular information, answer questions, and respond to any community member concerns to build ongoing support from the community.

Continued development

Within health settings, program sustainability is contingent on the ability of the program to undergo continued development and adaptation (Fleiszer *et al.*, 2015), and this has been critical in sustaining the ABLE-D model post-implementation. The model is continually updated and assessed to accommodate new ideas and societal and organizational change (e.g., integration of new technologies) to meet the changing needs of residents, which reflects the circular rather than linear model presented in Figure 7.2. The residential aged care manager, who has primary responsibility for ensuring the ABLE-D model continues to grow, regularly attends conferences, and conducts site visits to other facilities to source new ideas to inform the further development of the model. She also holds regular refresher training activities for new and existing team members and families. However, a key challenge to assessing the continued success of the model is the increasing needs of people living in the MSU. As the condition progresses and a person is no longer mobile, they are moved into an area which offers more sensory stimulation.

Research has also noted that broader uptake of rural health service innovation has been severely limited by reduced opportunities for dissemination and limited evaluation of programs (Asthana and Halliday, 2004; Wakerman and Humphreys, 2011). Consequently, RNH has placed considerable emphasis on disseminating the ABLE-D model so that it can be adopted in other contexts. The model has been presented at numerous health and aged care conferences both in Australia and internationally. Collaborations with universities, which have developed through these conference presentations, have led to greater conceptual clarity around the components of the model, which has enabled greater clarity in knowledge translation. These collaborations have also prompted a formal evaluation of the initial model (Roberts *et al.*, 2015). Promotion of the innovative aspects of the model during formal accreditation processes and site visits by regulatory bodies have led to the ABLE-D model being highlighted as an example of best-practice care by the national Australian Aged Care Quality Agency. This has prompted considerable interest in visiting RNH to observe the model in practice, and RNH regularly conducts tours of the MSU for senior leaders of other rural health services who are in the position to effect change within their organization (e.g., CEOs, directors of nursing, and lead managers). A YouTube video has also been produced by RNH's media liaison that introduces the ABLE-D model, which can be accessed by the public (www.youtube.com/watch?v=1LCRrcxlrXE).

Conclusion

This chapter has presented an innovative model of dementia care provision within a small, remote rural health care service which aims to provide best-practice, person-centred care within a resource-constrained environment. In particular, it highlights some of the strengths associated with the implementation of innovative models of care in rural contexts, including the ability to harness existing resources, collaborate with diverse service providers, and harness community buy-in for the purposes of resourcing and public support (Ryser and Halseth, 2014; Winterton *et al.*, 2014). From the discussion presented here, though, it is clear that the continued sustain-ability of the model is associated with the capacity and desire of RNH's senior management to take ownership of the model, and the subsequent capacity to resource this model. However, this top-down approach has been supplemented by the inclusion of team members, families, and other community or external organiza-tions at various stages of implementation. The emphasis on local representation within the RNH board of management has also been critical in driving the model and promoting it within the community. Key community organizations such as Rotary, Probus, and local church groups are now aware of, and acknowledge, the program as a positive element of the health service. This is best demonstrated by the increasing number of community members who are identifying older community members who may be at risk or require memory support, and are actively seeking to forge connections with RNH and the MSU team. This presents a key opportunity for early identification, diagnosis, and intervention.

This chapter also provides an innovative perspective on the evolving nature of rural health service models to meet community and resident needs. As rural social sustainability researchers have noted, the sustainability of rural social care systems is contingent on their ability to adapt to changing environmental and demographic characteristics (Black, 2005; Scott *et al.*, 2000). This work has highlighted some ways that small rural health services are adapting their models of care, and highlighted the role of flexible health funding systems to support innovation in health and aged care delivery. However, it has also identified some critical pressure points for change adaptation, including capacity to sustain organizational leadership, and the capacity for federal and state regulatory requirements to restrict innovation. This includes an emphasis on aged care funding models that are based on disability rather than ability; and risk-adverse clinical guidelines within health facilities that impact negatively upon a person-centred approach to care and achieving purpose and meaning for people living with dementia. Importantly, it also provides insight into how a small rural health service has managed to effectively translate and promote their model so that it can be implemented in other contexts, which has been identified as a significant limitation in relation to identifying innovative models of rural health care (Asthana and Halliday, 2004). Understanding how this can occur is critical in ensuring that rural service innovation can be shared across local, provincial, and national boundaries, and ensure quality of care for diverse rural people in diverse rural places.

References

Alzheimer's Disease International. 2015. *World Alzheimer report 2015: The global impact of dementia - An analysis of prevalence, incidence, cost, and trends*. London: Alzheimer's Disease International.

Asthana, S. and Halliday, J. 2004. "What can rural agencies do to address the additional costs of rural services? A typology of rural service innovation", *Health & Social Care in the Community* 12: 457–465.

Australian Government. 2016. *Financial issues affecting rural and remote aged care providers*. Canberra: Aged Care Financing Authority.

Australian Institute of Health and Welfare. 2016. *Australia's health 2016*. Canberra: Australian Government.

Baldwin, R., Stephens, M., Sharp, D., and Kelly, J. 2013. *Issues facing aged care services in rural and remote Australia*. Melbourne: Aged and Community Services Australia.

Barrett, A., Terry, D.R., Lê, Q., and Hoang, H. 2016. "Factors influencing community nursing roles and health service provision in rural areas: A review of literature", *Contemporary Nurse* 52: 119–135.

Black, A. 2005. "Rural communities and sustainability". In C. Cocklin and J. Dibden (eds.), *Sustainability and change in rural Australia* (pp. 20–37). Sydney: University of New South Wales Press.

Catic, A.G., Mattison, M.L., Bakaev, I., Morgan, M., Monti, S.M., and Lipsitz, L. 2014. "Echo-age: An innovative model of geriatric care for long-term care residents with dementia and behavioral issues", *Journal of the American Medical Directors Association* 15: 938–942.

Christi, B., Harris, M., Jayasinghe, U., Proudfoot, J., Taggart, J., and Tan, J. 2010. "Readiness for organisational change among general practice staff", *BMJ Quality & Safety* 19: 1–4.

Dal Bello-Haas, V.P., Cammer, A., Morgan, D., Stewart, N., and Kosteniuk, J. 2014. "Rural and remote dementia care challenges and needs: Perspectives of formal and informal care providers residing in Saskatchewan, Canada", *Rural and Remote Health* 14: 2747.

Department of Health. 2017. *Modified Monash Model* [Online]. Available online at www.health.gov.au/internet/main/publishing.nsf/content/modified-monash-model. Accessed 6 December 2017.

Department of Health and Human Services. 2016. *Department of Health and Human Services policy and funding guidelines 2016: Volume 2* (Health Operations 2016–17). Melbourne: State of Victoria.

Duckett, S. and Breadon, P. 2014. *Controlling costly care: A billion-dollar hospital opportunity.* Melbourne: Grattan Institute.

Edvardsson, D., Fetherstonhaugh, D., and Nay, R. 2010. "Promoting a continuation of self and normality: Person-centred care as described by people with dementia, their family members, and aged care staff", *Journal of Clinical Nursing* 19: 2611–2618.

Fine, M. and Yeatman, A. 2009. "Care for the self: 'Community aged care packages'". In A. Yeatman, G. Dowsett, M. Fine, and D. Gursansky (eds.), *Individualization and the delivery of welfare services* (pp. 165–186). London: Palgrave Macmillan.

Fleiszer, A.R., Semenic, S.E., Ritchie, J.A., Richer, M.C., and Denis, J.L. 2015. "An organizational perspective on the long-term sustainability of a nursing best practice guidelines program: A case study", *BMC Health Services Research* 15: 535.

Glasgow, N. and Brown, D.L. 2012. "Rural ageing in the United States: Trends and contexts", *Journal of Rural Studies* 28: 422–431.

Hansen, E., Robinson, A., Mudge, P., and Crack, G. 2005. "Barriers to the provision of care for people with dementia and their carers in a rural community", *Australian Journal of Primary Health* 11: 72–79.

Henderson, J., Willis, E., Xiao, L., Toffoli, L., and Verrall, C. 2016. "Nurses' perceptions of the impact of the aged care reform on services for residents in multi-purpose services and residential aged care facilities in rural Australia", *Australasian Journal on Ageing* 35: E18–E23.

Herron, R.V. and Rosenberg, M.W. 2017. "'Not there yet': Examining community support from the perspective of people with dementia and their partners in care", *Social Science & Medicine* 173: 81–87.

Kim, J., Young, L., Bekmuratova, S., Schober, D.J., Wang, H., Roy, S., Bhuyan, S.S., Schumaker, A., and Chen, L.W. 2017. "Promoting colorectal cancer screening through a new model of delivering rural primary care in the USA: A qualitative study", *Rural and Remote Health* 17: 4187.

Morgan, D.G., Kosteniuk, J.G., Stewart, N.J., O'Connell, M.E., Kirk, A., Crossley, M., Dal Bello-Haas, V.P., Forbes, D., and Innes, A. 2015. "Availability and primary health care orientation of dementia-related services in rural Saskatchewan, Canada", *Home Health Care Services Quarterly* 34: 137–158.

Morgan, D.G., Stewart, N.J., D'arcy, C., Forbes, D., and Lawson, J. 2005. "Work stress and physical assault of nursing aides in rural nursing homes with and without dementia special care units", *Journal of Psychiatric and Mental Health Nursing* 12: 347–358.

Moyle, W., Venturato, L., Cooke, M., Murfield, J., Griffiths, S.M., Hughes, J., and Wolf, N. 2016. "Evaluating the capabilities model of dementia care: A non-randomized controlled trial exploring resident quality of life and care staff attitudes and experiences", *International Psychogeriatrics* 28: 1091–1100.

ProfileID. 2017. *Yarriambiack Shire* [Online]. Available online at: http://profile.id.com.au/wimmera-region/industries?WebID=130&BMID=20. Accessed 3 December 2017.

Roberts, G., Morley, C., Walters, W., Malta, S., and Doyle, C. 2015. "Caring for people with dementia in residential aged care: Successes with a composite person-centered care model featuring Montessori-based activities", *Geriatric Nursing* 36: 106–110.

Ryser, L. and Halseth, G. 2014. "On the edge in rural Canada: The changing capacity and role of the voluntary sector", *Canadian Journal of Nonprofit and Social Economy Research* 5: 41–56.

Saleh, N., Penning, M., Cloutier, D., Mallidou, A., Nuernberger, K., and Taylor, D. 2017. "Social engagement and antipsychotic use in addressing the behavioral and psychological symptoms of dementia in long-term care facilities", *Canadian Journal of Nursing Research* 49: 144–152.

Scott, K., Park, J., and Cocklin, C. 2000. "From 'sustainable rural communities' to 'social sustainability': Giving voice to diversity in Mangakahia Valley, New Zealand", *Journal of Rural Studies* 16: 433–446.

Sheppard, C.L., McArthur, C., and Hitzig, S.L. 2016. "A systematic review of Montessori-based activities for persons with dementia", *Journal of the American Medical Directors Association* 17: 117–122.

Singh, R., Mathiassen, L., Stachura, M.E., and Astapova, E.V. 2010. "Sustainable rural telehealth innovation: A public health case study", *Health Services Research* 45: 985–1004.

Skinner, M. and Hanlon, N. 2016. *Ageing resource communities: New frontiers of rural population change, community development, and voluntarism.* New York: Routledge.

Stewart, N.J., Morgan, D.G., Karunanayake, C.P., Wickenhauser, J.P., Cammer, A., Minish, D., and O'Connell, M.E. 2016. "Rural caregivers for a family member with dementia: Models of burden and distress differ for women and men", *Journal of Applied Gerontology* 35: 150–178.

Talbot, R. and Brewer, G. 2016. "Care assistant experiences of dementia care in long-term nursing and residential care environments", *Dementia* 15: 1737–1754.

Vines, A.I., Hunter, J.C., White, B.S., and Richmond, A.N. 2016. "Building capacity in a rural North Carolina community to address prostate health using a lay health advisor model", *Health Promotion Practice* 17: 364–372.

Wakerman, J. and Humphreys, J. 2011. "Sustainable primary health care services in rural and remote areas: Innovation and evidence", *Australian Journal of Rural Health* 19: 118–124.

Warburton, J., Moore, M.L., Clune, S.J., and Hodgkin, S.P. 2014. "Extrinsic and intrinsic factors impacting on the retention of older rural healthcare workers in the north Victorian public sector: A qualitative study", *Rural and Remote Health* 14: 2721.

Winterton, R., Warburton, J., Clune, S., and Martin, J. 2014. "Building community and organisational capacity to enable social participation for ageing Australian rural populations: A resource-based perspective", *Ageing International* 39: 163–179.

8 Philanthropic organizations to the rescue? Alternative funding solutions for rural sustainability

Ryan Gibson and Joshua Barrett

Introduction

The trials and tribulations of infrastructure and service delivery in rural communities are well documented in Canada. Rural communities struggle to maintain the current suite of services in light of policy changes (Halseth and Ryser, 2006; Hanlon and Halseth, 2005; Kulig and Williams, 2012; Ryan-Nicholls, 2004). What if communities had an alternative source of funds to facilitate service delivery? Could these funds compensate for the challenges confronted related to large distances and low densities? Would the chronic challenges of service-withdrawal from the abdication of provincial and federal governments be overcome? How would access to alternative funding change the dynamics of rural sustainability, if at all?

This chapter explores the central question of whether philanthropic organizations can provide stability of service provision in rural communities. This chapter is not about intentional, long-term arrangements for service provision in rural communities. Rather, this chapter examines philanthropic organizations who take on service provision responsibilities in their communities to ensure access to programs and sustain a high quality of life, often after governments have discontinued their support. Through the examination of two rural community foundations, Sussex Area Community Foundation (New Brunswick) and Virden Area Foundation (Manitoba), this chapter illustrates how these philanthropic organizations transitioned into financing service provisions to their rural regions. This transition has been facilitated through the withdrawal of public investment, the retreat of service delivery, and the discontinuance of non-profit organizations. Although neither philanthropic organization intentionally strived to provide service delivery, they have recognized the critical importance of these services in building sustainable communities. The experiences of Sussex and Virden illustrate the potential opportunity for utilizing philanthropic funds as an alternative for service provision and rural sustainability.

Before examining the role of philanthropic organizations as service providers in rural communities, this chapter provides a brief discussion of the ebbs and flows of national policy for service provision. Building on that foundation, the chapter continues with a discussion of charities in Canada, including their role and structure and how they operate in rural communities. Based on the experiences from Sussex and Virden, four key lessons emerge. These lessons are integral to understanding the nexus between charities and service provision in rural communities in Canada.

The changing national policy context for rural service provision in Canada

The history of rural development policy in Canada has been influenced by multiple approaches that have impacted economic, social, and political systems. The post-war period of rural development policies was driven by Keynesian and interventionist approaches. This was a period in which governments were intrinsically engaged in the main street of rural communities. These approaches were later replaced by neoliberal approaches, whereby New Public Management practices emerged (Aucoin, 1995). During this period, the role of the state fundamentally shifted by adopting corporate practices of effectiveness and efficiency into the public service. These changes in policy approaches have profoundly impacted rural communities throughout the country.

In the immediate aftermath of the Second World War, the role of the central government expanded dramatically in the areas of economic development and social service provision. Guided by the assumption that economic and social goals could be separated, the implementation of Keynesian approaches led to federal policies focused on infrastructure, social programs, and the use of Crown corporations to stimulate economic development (Fairbairn, 1998). Under these Keynesian approaches, the government was solely responsible for public service provision. Guided by the tenets of Keynesian approaches, new initiatives, investments, and programs increased the presence of government on the main streets of rural communities (Halseth, 2003; Polèse, 1999).

The adoption of neoliberalist policies in the 1980s and 1990s represented a retreat of government from rural communities. This withdrawal represented a clear movement away from previous Keynesian and interventionist approaches of the 1960s–1980s (Markey *et al.*, 2008). The movement from the Keynesian approaches towards neoliberalist approaches resulted in considerable impacts on charities in rural communities (Douglas, 2005). The adoption of neoliberal policies created an environment wherein government either sought new service arrangements or withdrew services altogether (Evans and Shields, 2006; Markey *et al.*, 2015). The movement towards neoliberalist approaches was not reserved only for urban areas; as Young and Matthews (2007) note, rural communities equally witnessed this shift.

The transition to neoliberalist approaches facilitated the adoption of the New Public Management and the corresponding decline of Keynesian economic policy (de Vries, 2010). The tenets of New Public Management focused on (i) building on lessons from the private sector through the increased roles of the markets and corporate management techniques (Bevir, 2009; Osborne and Gaebler, 1993), (ii) entrepreneurial leadership within the public services, (iii) explicit standards and measures of performance (Hood, 1991), (iv) an enhanced focus on outputs and results and a diminished focus on processes and procedures (Hood, 1991; Osborne, 2010), and (v) the creation of financial efficiencies through competitive market forces. The implications of New Public Management policies, programs, and strategies have had critical impacts on rural communities.

In the rural Canadian context, the adoption of neoliberalist approaches has translated into the federal and provincial governments removing themselves from public service provisions. Often this removal is described as 'decentralization', 'downloading', or 'offloading' (Evans and Shields, 2006). Government decisions to retract their 'footprint' had substantial impacts on peripheral communities (Douglas, 2005). The visible impacts of the retreat include the removal and loss of railways, post offices, schools, government jobs, and internet access/networking points (Epp and Whitson, 2006). The less visible impacts of the retreat focus on dismantling institutions created to support rural development (Reimer and Markey, 2014; Young, 2008). The retreat from the periphery by government and the dismantling of rural institutions leave many communities at a loss for how to move forward.

Communities across the country have witnessed a series of cutbacks to public service provision. In response to the cutbacks, offloading, and downloading, communities are exploring alternative mechanisms to ensure the continuation of service delivery. Increasingly, rural communities are turning to charities and the non-profit sector to step into the role of service providers – a role that was once played by government (Austin, 2003; Chouinard and Crooks, 2008; Evans and Shields, 2006).

Intersection of charities and rural communities

Throughout Canada, there are over 18,000 rural charities,[1] each engaged in unique activities depending on the local and regional priorities. Although unique, all charities in Canada are required to follow the same regulations outlined in the federal *Income Tax Act*. This section briefly outlines the definition of a charity and the role of charities in rural Canada. This section also examines the implications emerging from the movement away from Keynesian approaches to neoliberalism and the New Public Management. This shift holds substantive implications for charities, especially in rural communities.

Understanding charities and their role in rural Canada

Charities are legal organizations registered with the Canada Revenue Agency under the *Income Tax Act*. All charities must undertake activities in at least one of the four primary arenas: (i) relief of poverty, (ii) advancement of education, (iii) advancement of religion, or (iv) other purposes beneficial to the community (Levasseur, 2012). According to the *Income Tax Act*, all charities need to meet the public benefit test – determining whether the activities of the charity are of practical utility or of benefit to the larger community (Canada Revenue Agency, 2006; Quarter *et al.*, 2017). To encourage Canadians to financially contribute to charities, the Government of Canada provides a tax incentive. Since 1930, donations made to registered charities are eligible for a tax deduction for the donor (Elson, 2007; McCamus, 1996). This incentive allows charities to collect funds, which are invested into the activities of the charity.

Charities in Canada can be designated as either private or public, depending on the source of funds and the arm's-length nature of their governance. Public foundations receive funds from multiple sources and more than half of their governance board is at arm's length, while private foundations receive more than half of their funding from an individual and/or family and less than half of their governance board is at arm's length (Canada Revenue Agency, 2017b). Public foundations typically are accountable to the community or region they represent. Public foundations strive to build endowments, which can be invested, and interest generated through the investments can be re-invested in the community through grants. This chapter focuses on only public foundations in rural communities in Canada.

Across the country, in both rural and urban communities, Canadians are actively involved with charities. In 2015, 82% of Canadians over the age of 15 donated to at least one charity and a total of $12.7 million was donated by Canadians to charities (Turcotte, 2015). Although the total number of donors was substantially lower than in 2013, the total value of donations was higher.

Public foundations in rural communities are important actors, albeit often over-looked, when discussing rural development. Charities have played, and continue to play, an important role in contributing to the sustainability and revitalization of rural areas. Throughout Canada, there are over 18,000 rural charities and numerous non-profit organizations influencing the economic, social, cultural, and environmental well-being of their respective communities (Canada Revenue Agency, 2008). Recent economic restructuring and the changing role of the state over the past decades have resulted in a heightened role for charities and non-profit organizations.

The 2008 analysis by the Canada Revenue Agency, the most recent analysis conducted to date, noted that rural charities increasingly relied on volunteers to operate. Few rural charities employed full-time staff to support their mandate and activities (Canada Revenue Agency, 2008). Given the low population density and higher distances to larger centres, the Agency noted, "rural charities have a more significant presence in the minds of community residents than might be the case in larger centres where business or government organizations may dominate" (Canada Revenue Agency, 2008, p. 9). It is interesting to note that while approximately 17% of Canadians live in rural communities, 22% of charities are based in rural communities.

Since the 1990s, the charitable sector in Canada has witnessed tremendous growth. Emmett and Emmett (2015) note this growth is driven by demand for services the charitable sector can provide. This growth, however, comes with increased pressures to deliver programs and activities while at the same time balancing limited volunteer capacity and burnout, a continually changing policy environment, and increased offloading from all levels of government (Hall and Reed, 1998).

Influence of changing rural development policies on charities

Over the past 150 years, the relationship between charities and the Government of Canada has ebbed and flowed (Elson, 2007). The continued economic restructuring and policy changes, as discussed earlier, create new roles for rural charities. As locally based actors, charities are well positioned to identify local needs that are no longer being provided by government and to create place-based strategies. Too often, however, rural charities are the only local funding organizations to support local initiatives.

The economic restructuring, which continues today, and periods of fiscal austerity by provincial/territorial and federal governments saw a shift in responsibilities. Many responsibilities moved downward from the provincial/territorial and federal levels to local communities (Hall and Reed, 1998). Larger communities with high human capital often could adjust to these shifts but rural communities struggled. Combined with these new responsibilities, rural communities also witnessed a retreat by government, whereby many service provisions were discontinued. This strategy often meant rural communities lost key health, education, and social services. This phenomenon of offloading, stretches the limited human and financial capacities of rural communities (Douglas, 2005).

As the amount of local responsibilities has increased and provincial/territorial and federal governments have moved away from rural service provision, charities have increasingly filled these gaps. Halseth and Ryser (2007) noted charities generate place-based solutions to these reductions in service. As organizations rooted in their community, charities strive to ensure that community health and well-being are not compromised. In many rural communities, charities are now providing services that used to be the responsibility of government. Recent research noted that some charities believe they are 'picking up the pieces after the departure of government' (Gibson *et al.*, 2014). Given the current environment, it is likely that rural charities will continue to serve their community in this manner.

Place-based philanthropy has emerged as a new tool for addressing long-term community planning (Glücker and Pies, 2011; Jung *et al.*, 2013). The hallmarks of place-based philanthropy include (i) meaningful engagement of local residents in planning, programming, and prioritizing opportunities; (ii) long-term commitment to the place; (iii) focus on endowment building to support the place; and (iv) recognizing the interconnected nature of challenges and opportunities in the place (Aspen Institute, 2015; Murdoch *et al.*, 2007). The ability of rural-based charities to understand their community and its priorities facilitates place-relevant priorities and programming.

Community foundations are one type of place-based philanthropy organization. Community foundations are registered with the Canada Revenue Agency as charities. Throughout Canada, 83 community foundations are based in rural communities, representing 43% of all community foundations (Gibson *et al.*, 2014). It should be noted that a number of urban-based community foundations provide financial grants to surrounding rural communities, increasing the number of rural communities being served by a community foundation. The first rural community foundation was started in 1974; however, most rural community foundations were established in the 1990s–2000s. Despite being a relatively new vehicle for rural development, rural community foundations have over $114 million in endowments to support local priorities. Although robust financial endowments are in place, continued growth may be jeopardized by rural economic restructuring. With their endowments, rural community foundations are taking innovative approaches to supporting their communities.

From the front lines: engagement of rural charities with service provision

To understand the interconnection of charities and rural development, a series of interviews was conducted with rural charities across Canada. A series of interviews took place with board members and staff of rural charities and focused on understanding how, if at all, the rural charity contributed towards a vibrant, sustainable rural future. Based on those interviews, the narratives from Sussex, New Brunswick, and Virden, Manitoba, exemplify how rural charities are becoming engaged in service provision (see Figure 8.1).

Sussex Area Community Foundation

Sussex Area Community Foundation was established in 2004 and is based in the community of Sussex, New Brunswick. Sussex Area Community Foundation has been dedicated to serving the people within the communities of Sussex Corner, Apohaqui, Norton, Sussex, and other surrounding areas, helping a population of approximately 8,099 residents (Statistics Canada, 2016a). The community foundation is governed by an appointed board of 12 local residents representing a diversity of backgrounds. The board is appointed by a nominating committee consisting of the local mayors, service club presidents, and the local chamber of commerce.

The Sussex Area Community Foundation holds an endowment of $2.1 million. The assets of the foundation are the result of both individual and corporate donations since its creation in 2004. In 2015–2016, the Sussex Area Community Foundation provided $89,627 in gifts to the region (Canada Revenue Agency, 2017a), with a variety of granting priorities focused on health and well-being, education, arts and culture, immigration and refugees, and school programming.

Over the past decade, the Sussex Area Community Foundation has moved in the direction of financially supporting public services to residents of their region that were formally the responsibility of provincial and/or federal governments. The continued pattern of service discontinuance by all levels of government is a concern for the board of directors of the Sussex Area Community Foundation.

Fiscal austerity at all levels of government has created a need for governments to adjust their budgets. As was noted by one board member, "this realignment is a lot of the time at the detriment of social groups that need help the most" (Interview Participant #12, 2014). Services and projects that were in the past

Figure 8.1 Case studies.
Map credit: Kyle Kusch

completely supported by government now have to find their own dollars to deliver needed activities in the community. Mental health support service is one example in the Sussex area. Since 2012, the Sussex Area Community Foundation has provided substantial annual financial support for mental health support services in the community. As part of the community foundation's annual grants, financial grants were provided to a non-profit organization who in turn performed services in the community. The community foundation's grant was leveraged through other sources of funding to ensure the service provision remained in the community. Without this financial support, these services would have otherwise discontinued, forcing residents to travel outside the region for access.

A similar story takes place regarding home care supports for seniors in the region. The Sussex Area Community Foundation annually provides financial resources to local organizations to provide home care visits, foot care, and hospice care. Residents requiring these services are able to remain in the community. These services used to be the purview of either the provincial or federal government, and the continuance of these programs in the Sussex area would not be possible without the Sussex Area Community Foundation. The foundation also provides financial support to key community infrastructure. The Sussex Area Community Foundation has funded necessary hospital equipment for the regional health facility, purchased playground and school equipment for elementary and high schools, supported community performance spaces, and provided contributions for building renovations to accommodate different motilities.

As the economic restructuring, decline of rural institutions, and government offloading continue, board members from the Sussex Area Community Foundation noted they are perceived as 'filling in the funding gaps left when government stopped supporting these services'. Filling these funding gaps is an important role rural charities provide their communities, which in turn demonstrates their impact and value. The board members explained they are proud to be able to support these services, though they expressed disappointment these initiatives are no longer the responsibility of the government. Concern is also raised over the potential for continued offloading and increasing pressures on the Sussex Area Community Foundation to respond.

In these examples, the Sussex Area Community Foundation did not provide the public services. Rather, the Sussex Area Community Foundation allocated annual funds to supporting other organizations to provide the services. The financial resources of the Sussex Area Community Foundation were not envisioned to support the withdrawal of government. Every dollar invested into continuing government services means a dollar not spent on other local and regional priorities.

Virden Area Foundation

Located in southwestern Manitoba, the Virden Area Foundation was incorporated in 1992 as a charity to serve the rural community of Virden and the surrounding rural municipality of Wallace-Woodworth. The Virden Area Foundation is a public charity serving a population of 6,270 (Statistics Canada, 2016b). In the late 1980s, a local business family bequeathed $600,000 to the Town of Virden to support a community project or purpose of long-lasting nature. The estate was transferred to the Virden Area Foundation upon its creation. Since this time, the endowment of the Virden Area Foundation has grown through individual and corporate donations.

The Virden Area Foundation is governed by a board of directors, each member being part of the community and active in community life. The board is appointed by a nominating committee consisting of the local mayors or reeves, a local member of the provincial parliament, the chair of the regional school board, and the president of the local chamber of commerce.

As of 2016, the Virden Area Foundation held endowments valued at $1.9 million (Canada Revenue Agency, 2017a). Based on this endowment, the Virden Area Foundation provided gifts of $47,210 to support a variety of local priorities such as the local library, post-secondary scholarships, minor sports, youth activities, the local museums, and heritage building restoration (Virden Area Foundation, 2017).

The Virden Area Foundation is engaged in a number of ways to help strengthen the residents of their region. Some of their grants are designated towards restoring heritage by upgrading their local museums. Other priorities include distributing scholarships for high school students and providing funding for new health services. In 2009, the Virden Area Foundation contributed $150,000 towards building a new recreational complex. Additionally, they have provided financial support for students and clubs to help improve access to recreation (Virden Area Foundation, 2017). In both instances, the foundation has collaborated with other funding agencies to deliver the outcome.

The Virden Area Foundation is currently involved in a couple of unique public service provisions. Confronted with declining and aging volunteers and decreased funding opportunities from government, community-based non-profit organizations in the area have turned to the Virden Area Foundation. Due to their circumstances, these community-based organizations are no longer in a position to continue their mandate and activities. As a recognized community leader, the Virden Area Foundation has been approached to take over the service provisions once led by these organizations. Upon the dissolution of the non-profit organization, the Virden Area Foundation assumes the mandate, activities, and any residual financial assets. This scenario most recently played out with a non-profit organization with a mandate to provide services to the local cemeteries. As the local non-profit organization was not able to continue its operation due to the lack of volunteers, it donated its limited financial assets to the Virden Area Foundation who, in turn, took over the mandate and activities previously conducted by the non-profit organization. It is important to note that the inability to continue the non-profit organization's activities does not reflect the lack of need for their services in the community and region. Rather, the pressures being felt by non-profit organizations from the 'disappearance of government' are beyond the voluntary and financial capacity of local organizations.

A second example of the Virden Area Foundation taking on a new role in service provision in the region revolves around early childhood education. Working with other community-based stakeholders, the Virden Area Foundation played a leadership role in leading initiatives to build a new early childhood education facility. Historically, the expenses for early childhood education were the responsibility of the provincial government. The board of directors for the Virden Area Foundation recognizes the importance of early childhood education from the perspectives of child development, health and well-being, and economic development. The Virden Area Foundation committed substantial financial resources towards the development of a new early childhood education facility in the community. Had the provincial government continued to provide these public services, the Virden Area Foundation would have been able to allocate their funds towards other deserving local initiatives.

This pattern of providing substantial financial support to infrastructure projects that used to be the purview of the provincial government is a growing concern for board members of the Virden Area Foundation. The Virden Area Foundation has provided financial support to community recreational facilities, school equipment, and now an early childhood education facility. One board member described this pattern as an 'uneasy path with no indication of change ahead', which was problematic as the Virden Area Foundation does not have the financial resources to continue this 'uneasy path' over the long term. A second board member noted that without the Virden Area Foundation's financial support, many of these service initiatives would not happen in their community. Although board members find it 'uneasy' to fund initiatives that were previously the mandate of the provincial government, they are not prepared to leave their community without these services.

Moving forward

Since the 1990s, communities across Canada have witnessed dramatic changes in the role of government to support and facilitate rural development. These changes have increasingly influenced rural charities as well, as seen in the cases of Sussex and Virden. The experiences of the Sussex Area Community Foundation and the Virden Area Foundation speak to the robustness, nimbleness, and dedication of rural charities towards their respective communities. Their stories of stepping up to take on new financial responsibilities upon government decisions to discontinue rural service provision are not unique to Sussex and Virden.

From these stories, four key lessons emerge for rural community leaders, government leaders, and philanthropic organizations. First, there is a cautionary note of 'slippery slopes' in both narratives. Although rural charities in both communities stepped up to ensure the continuation of service provision, there is a strong sense of concern and alarm that governments are being let 'off the hook'. Governments are not being held accountable for service provision. One of the hallmarks of place-based philanthropy is collaboration among all levels of actors. There is a distinct lack of collaboration with governments in stories from Sussex and Virden. In both communities, there is strong evidence of collaboration with other local and regional actors; however, government participation is noticeably missing. The absence of government in these initiatives is counterproductive and undermines place-based philanthropy. Rather than waiting for government participation, these communities are finding resilience through rural charities to overcome potential loss of service provision.

At the same time, a number of board members noted caution. One board member noted that once rural charities start supporting former government initiatives two challenges emerge. First, at the community level, these charities require annual funding to perform their valuable services and programs. In essence, rural charities could become 'locked' into continued support for the initiative, of which there is a clear need in the community. If the endowment of community foundation remains static or interest rates of investments decline, these organizations may struggle to support all local initiatives. Second, at the government level, is a concern that by supporting these valuable initiatives rural charities are complacent in the government's strategy of 'offloading' financial responsibilities to the local level. The concern is that the pattern of 'offloading' would continue with other programs and services in rural areas.

The second lesson is that there is precariousness to the ability of rural charities to absorb the service provision role. Taking on these roles, often unexpected, results in real costs to each philanthropic organization. Spending financial and human resources to continue, and sometimes re-establish, rural service provisions comes at the expense of supporting other local priorities. It was suggested that the support of organizations providing programs and services that were once the responsibility of government hinders the ability of community foundations to be forward thinking. By 'picking up the pieces' after the departure of government, rural charities become pivotal in maintaining the current suite of services and programs through financial support. This financial support, in turn, limits the funds available for rural charities to be proactive in addressing long-term community priorities.

Third, if we continue to rely on rural charities to 'pick up the pieces' from government abdication, we need to re-examine how philanthropic organizations are supported. Government may need to look at mechanisms to encourage building endowments and donations to build the financial capacity of rural charities, which in turn can be utilized to support service provisions in their communities and regions. These mechanisms may include tax incentives for donating to endowments or allowing philanthropic organizations to be eligible recipients of more government grants and programs. The Smart and Caring Communities initiatives of former Governor General Johnston shed light on charities and their value to communities and to Canada (Government of Canada, 2017); however, it cannot be the only government mechanism utilized. Simply continuing to have rural charities increase their roles in service provision without a financial strategy is recognized as short-sighted by board members in both the Sussex Area Community Foundation and Virden Area Foundation. Although rural charities are place-based institutions and often have financial resources, Hall and Reed (1998) remind us that charities are not always good mechanisms for transferring government services. Their research illustrates the inability of charities to raise consistent annual financial resources to provide services in the long term.

Fourth, proactive strategic discussions are required regarding future service provision changes by all levels of government and community stakeholders. Too often rural charities are caught off guard by government decisions to discontinue service provisions. As a result, rural charities often have to turn to reactionary responses to ensure the continuation of services. Collaborative engagements with all stakeholders about the future of service provision would allow purposeful planning for the long term and more effective transitions.

What is clear from the Sussex Area Community Foundation and Virden Area Foundation is that rural charities are increasingly engaged in service provision in rural communities. With a strong connection and commitment to place, rural charities are uniquely positioned to offset the offloading of service provision by governments due to their financial resources. These financial resources can be utilized to ensure the continuation of rural services. What is unknown is the ability of rural charities to continue this role as a service provider in the long term, particularly if additional services are offloaded. There will be a threshold at which the financial capacity of rural charities will be met or exceeded. Given the diversity of rural charities and their financial resources, this threshold will likely be reached at a different time in each community.

Acknowledgements

This research has been supported through research grants from the Leslie Harris Centre for Regional Policy and Development at Memorial University of Newfoundland, Institute of Nonprofit Studies at Mount Royal University, and the Social Sciences and Humanities Research Council of Canada.

Note

1 Based on the forward sortation area definition of rural. For more details on the forward sortation area definition of rural, see DuPlessis *et al.* (2002).

References

Aspen Institute. 2015. *Towards a better place: A conversation about promising practice in place-based philanthropy.* Washington, DC: Aspen Institute.
Aucoin, P. 1995. *The new public management: Canada in comparative perspective.* Montreal: Institute for Research in Public Policy.
Austin, M.J. 2003. "The changing relationship between nonprofit organizations and public social service agencies in the era of welfare reform", *Nonprofit and Voluntary Sector Quarterly* 32(1): 97–114.
Bevir, M. 2009. *Key concepts in governance.* London: Sage.
Canada Revenue Agency. 2006. *Guideline for registering a charity: Meeting the public benefit test.* Ottawa: Canada Revenue Agency.
Canada Revenue Agency. 2008. *Small and rural charities: Making a difference for Canadians 2008.* Ottawa: Canada Revenue Agency.
Canada Revenue Agency. 2017a. List of charities and other qualified donees. Available online at www.canada.ca/en/revenue-agency/services/charities-giving/list-charities/list-charities-other-qualified-donees.html. Accessed 8 February 2018.
Canada Revenue Agency. 2017b. *Registering a charity for income tax purposes.* Ottawa: Canada Revenue Agency.
Chouinard, V. and Crooks, V.A. 2008. "Negotiating neoliberal environments in British Columbia and Ontario, Canada: Restructuring of state-voluntary sector relations and disability organizations' struggles to survive", *Environment and Planning C: Government and Policy* 26(1): 173–190.
de Vries, J. 2010. "Is new public management really dead?", *OECD Journal on Budgeting* 2010/1: 1–6.
Douglas, D. 2005. "The restructuring of local government in rural regions: A rural development perspective", *Journal of Rural Studies* 21(2): 231–246.
DuPlessis, V., Beshiri, R., Bollman, R., and Clemenson, H. 2002. *Definitions of rural.* Agriculture and Rural Working Paper Series, Working Paper No. 61. Ottawa: Statistics Canada.
Elson, P.R. 2007. "A short history of voluntary sector-government relations in Canada", *The Philanthropist* 21(1): 36–74.

Emmett, B. and Emmett, G. 2015. *Charities in Canada as an economic sector: Discussion paper.* Toronto: Imagine Canada.

Epp, R. and Whitson, D. 2006. "Introduction: Writing off rural communities?" In Epp, R. and Whitson, D. (eds.), *Writing off the rural west: Globalization, governments, and the transformation of rural communities* (pp. xiii–xxxv). Edmonton: University of Alberta Press.

Evans, B.M. and Shields, J. 2006. *Neoliberal restructuring and the third sector: Reshaping governance, civil society and local relations.* Toronto: Centre for Voluntary Sector Studies, School of Management, Ryerson University.

Fairbairn, B. 1998. *A preliminary history of rural development policy and programme in Canada, 1945-1995.* Montreal: New Rural Economy.

Gibson, R., Barrett, J., and Vodden, K. 2014. *Fostering sustainable futures for rural community and regions through philanthropy.* Corner Brook: Grenfell Campus, Memorial University.

Glücker, J. and Pies, M. 2011. "Why being there is not enough: Organized proximity in place-based philanthropy", *The Service Industries Journal* 32(4): 515–529.

Government of Canada. 2017. *Toward a smarter, more caring Canada.* Ottawa: Government of Canada.

Hall, M. and Reed, P. 1998. "Shifting the burden: How much can government download to the non-profit sector?" *Canadian Public Administration/Administration Publique Du Canada* 41(1): 1–20.

Halseth, G. 2003. "Attracting growth 'back' to an amenity rich fringe: Rural-urban fringe dynamics around metropolitan Vancouver, Canada", *Canadian Journal of Regional Science/Revue Canadienne Des Sciences Régionales* 3: 297–318.

Halseth, G. and Ryser, L. 2006. "Trends in service delivery: Examples from rural and small town Canada, 1998-2005", *Journal of Rural and Community Development* 1(2): 69–90.

Halseth, G. and Ryser, L. 2007. "The deployment of partnerships by the volutnary sector to address service needs in rural and small town Canada", *International Journal of Voluntary and Nonprofit Organizations* 18(3): 241–265.

Hanlon, N. and Halseth, G. 2005. "The greying of resource communities in northern British Columbia: Implications for health care delivery in already-underserviced communities", *The Canadian Geographer* 49(1): 1–24.

Hood, C. 1991. "A public management for all seasons?", *Public Administration* 69: 3–19.

Jung, T., Harrow, J., and Phillips, S. 2013. "Developing a better understanding of community foundations in the UK's localisms", *Policy & Politics* 43(3): 409–427.

Kulig, J. and Williams, A. (eds.). 2012. *Health in rural Canada.* Vancouver: UBC Press.

Levasseur, K. 2012. "In the name of charity: Institutional support for and resistence to redefining the meaning of charity in Canada", *Canadian Public Administration/Administration Publique Du Canada* 55(2): 181–202.

Markey, S., Breen, S., Gibson, R., Lauzon, A., Mealy, R., and Ryser, L. (eds.). 2015. *The state of rural Canada.* Camrose: Canadian Rural Revitalization Foundation/Fondation canadienne pour la revitalisation rurale.

Markey, S., Halseth, G., and Manson, D. 2008. "Challenging the inevitability of rural decline: Advancing the policy of place in northern British Columbia", *Journal of Rural Studies* 24(4): 409–421.

McCamus, J. 1996. *Report on the law of charities* (Vol. 1). Toronto: Ontario Law Reform Commission.

Murdoch, J., Garrigan, B., Lavin-Loucks, D., Murdock,J., III., Hess, L., and Thibos, M. 2007. *The place-based strategic philanthorpy.* Dallas: Center for Urban Economics.

Osborne, D. and Gaebler, T. 1993. *Reinventing government: How the entrepreneurial spirit is transforming the public sector.* New York: Plume.

Osborne, S. 2010. "The (new) public governance: A suitable case for treatment?" In Osborne, S. (ed.), *The new public governance? Emerging perspectives on the theory and practice of public governance* (pp. 1–16). London: Routledge.

Polèse, M. 1999. "From regional development to local development: On the life, death, and rebirth (?) of regional science as a policy relevant science", *Canadian Journal of Regional Science* 12(3): 299–314.

Quarter, J., Mook, L., and Armstrong, A. 2017. *Understanding the social economy: A Canadian perspective* (2nd ed.). Toronto: University of Toronto Press.

Reimer, B. and Markey, S. 2014. "Dismantling of rural institutions". In Canadian Rural Revitalizaton Foundation (ed), *2013-2014 Canadian rural revitalization foundation annual report.* Camrose: Canadian Rural Revitalization Foundation.

Ryan-Nicholls, K. 2004. "Health and sustainability of rural communities", *Rural and Remote Health: The International Electronic Journal of Rural and Remote Health Research, Education, and Policy* 4: 242.

Statistics Canada. 2016a. *Sussex, T [Census subdivision], New Brunswick and Norton, RM [Census subdivision].* Ottawa: Statistics Canada.

Statistics Canada. 2016b. *Virden, T [Census subdivision], Manitoba and Wallace-Woodworth, RM [Census subdivision].* Ottawa: Statistics Canada.

Turcotte, M. 2015. *Volunteering and charitable giving in Canada.* Ottawa: Statistics Canada.

Virden Area Foundation. 2017. *2015–2016 community update.* Virden: Virden Area Foundation.

Young, N. 2008. "Radical neoliberalism in British Columbia: Remaking rural geographies", *Canadian Journal of Sociology* 33(1): 1–36.

Young, N. and Matthews, R. 2007. "Resource economies and Neoliberal experimentation: The reform of industry and community in rural British Columbia", *Area* 39(2): 176–185.

Part IV

New infrastructure arrangements

9 Innovations for sustainable rural drinking water services

Sarah Minnes, Sarah-Patricia Breen, and Kelly Vodden

Introduction

Water is recognized as both a basic human right and a fundamental component of environmental sustainability and human development (UNESCO International Hydrological Programme, 2014). The management of drinking water from source to tap is a key facet of sustainable development. A key characteristic of sustainable communities is having a sustainable water system (Robinson *et al.*, 2008). The sustainable management of drinking water infrastructure plays a vital role in the delivery of safe and clean drinking water to residents, but beyond quality of life for residents, water systems also impact the economy and the environment (Breen and Markey, 2015). For example, boil water advisories can deter tourism, and the over-consumption of water can interfere with natural water systems needed to support aquatic life.

Both the built and natural infrastructure used for the collection, storage, treatment, and distribution of drinking water are considered to be critical public infrastructure (CBCL Limited, 2012; Federation of Canadian Municipalities, 2012). However, in rural Canada, there are significant challenges to drinking water service delivery in both public systems (i.e., owned and operated by local governments) and private systems (e.g., water user communities, individuals), flowing from a complex combination of factors. For example, built infrastructure has suffered from a lack of investment, leading to its subsequent degradation and resulting in an infrastructure deficit – a gap between what is needed and available to meet current needs (Burleton and Caranci, 2004; Kennedy *et al.*, 2008; Mirza, 2007). This infrastructure deficit is further exacerbated by contextual changes, both past (e.g., economic, political, legislative, and social change) and future (e.g., climate change), presenting additional challenges (Breen and Markey, 2015; CBCL Limited, 2012; Minnes and Vodden, 2017). In the context of the above, drinking water management is a serious challenge for rural Canada, one which current governance and management structures have struggled to address.

The concept of sustainable infrastructure is new and evolving, including infrastructure systems that are specifically designed to be sustainable (e.g., energy efficient), and the sustainable management of infrastructure systems (e.g., management to contribute to the current and future objectives of society while maintaining environmental integrity) (Pollalis *et al.*, 2012; Robinson *et al.*, 2008). Characteristics of sustainable management include adequate funding; sustainable government and management; and owners and users having appropriate knowledge and understanding of responsibilities, service levels, risks, and costs (Baldwin and Dixon, 2008; British Columbia Water and Waste Association, 2014). Specific to drinking water systems, sustainable infrastructure includes consideration of source water and ecosystem protection, along with understanding of the actual built infrastructure (Ministry of Community and Rural Development Local Government Infrastructure and Finance Division, 2010; Pollalis *et al.*, 2012; Santora and Wilson, 2008). When sustainable infrastructure is identified as a goal, what becomes apparent is the need to reconsider current approaches to planning, design, and management of infrastructure (CBCL Limited, 2012; Pollalis *et al.*, 2012).

Two key elements of sustainable drinking water systems, further discussed in this chapter, include i) asset management, including a complete and accurate inventory of existing infrastructure, its location and condition and ii) the ability to acquire and retain the appropriate human capacity needed to manage and maintain the systems (Coad, 2009; Maxwell, 2008; Santora and Wilson, 2008). However, these characteristics of sustainable infrastructure can be particularly challenging to establish in small drinking water systems, public or private, as a result of capacity challenges – specifically because those responsible for operating these systems often do not have the financial or human resources to invest in new infrastructure or to implement new approaches or programs (Minnes and Vodden, 2017). There is also a link between the low user charges typically associated with drinking water services in rural places, degrading infrastructure, and a lack of financial resources to switch to new approaches and technologies (Kot *et al.*, 2011). Rural and remote regions also face the additional challenges typical of rural areas, such as demographics (e.g., low population density, aging, out migration), lack of economies of scale, and physical isolation (Health Protection Branch, 2013; Kot *et al.*, 2011). Such challenges can substantially impact the amount of investment available for maintaining degrading infrastructure – including basic asset management (Kot *et al.*, 2011; Locke, 2011; Minnes and Vodden, 2017), as well as the ability to find and retain skilled human resources to maintain and manage these assets and other aspects of drinking water systems (Minnes and Vodden, 2017).

The purpose of this chapter is to discuss new approaches to public drinking water service delivery that address the challenges related to asset management and human capacity introduced above. The two examples provided demonstrate innovative, place-based solutions focused on tackling the infrastructure deficit and improving the sustainability of drinking water systems through i) improved asset management and ii) enhancing human capacity at community and regional-scale levels in rural and remote places. The first case study will speak to the role information technologies can have in improving 21st-century rural water services and water systems management, using an example of the utilization of asset management software in rural Newfoundland and Labrador (Daniels, 2014). The second case study will speak to the role of innovation and collaboration in improving the training and expanding the skill sets of water operators in rural areas exemplified by the Water Smart Peer-to-Peer Operator Training program from British Columbia (Breen, 2016). We begin by providing an overview of rural drinking water service delivery and related public policy in rural Canada. The chapter ends with an overarching discussion on the lessons demonstrated by these cases regarding the role innovations can play in the sustainable management of rural and remote drinking water systems, including those in asset management and human capacity.

Canadian drinking water service delivery policy and responsibilities

In Canada, the responsibility for ensuring the delivery of drinking water is shared between federal, provincial, and local governments – although the 'heavy lifting' is typically done at the local level. At the federal level, departments with water as a mandate include Environment Canada, Health Canada, Public Works and Government Services Canada, Aboriginal Affairs and Northern Development, and the Intergovernmental Affairs with the Privy Council Office. While Canada does not have a national-level federal drinking water strategy or law (Cook *et al.*, 2013), there are Guidelines for Canadian Drinking Water Quality (GCDWQ) which were created by Health Canada in collaboration with the provinces and territories, through the Federal-Provincial-Territorial Committee on Drinking Water (Minnes and Vodden, 2017). The GCDWQ are voluntarily used at all levels across Canadian jurisdictions as recommendations informing the provincial and territorial legislation and policy requirements for drinking water quality. However, it is not mandatory that provincial and territorial legislative and policy measures meet the GCDWQ standards (Kot *et al.*, 2011). Major federal-level pieces of legislation that can influence how water is managed include the *Canada Water Act*, the *Safe Drinking Water for First Nations Act*, the *National Parks Act*, the *Canadian Environmental Assessment Act*, the *Fisheries Act*, the *Canadian Environmental Protection Act*, and the *Navigation Protection Act*. Furthermore, Environment Canada has a *Federal Water Policy* (1987), a broad-based framework that does not supersede the authority of the provinces and territories (Canadian Council of Ministers of the Environment, 2004).

Provincial and territorial governments are responsible for providing safe drinking water, including infrastructure, source water protection, capacity building, water quality monitoring, and more, with the exception of those areas that are within federal jurisdiction (e.g., First Nations reserves and national parks) (Hill *et al.*, 2008; Ramalho *et al.*, 2014). However, while provincial and territorial governments retain powers related to setting and enforcing standards, regulation, and policy, they devolve most drinking water responsibilities to municipal and other local governments to deliver water services locally, as well as to operate and maintain drinking water systems (Canadian Council of Ministers of the Environment, 2004; Hill *et al.*, 2008).

Beyond the different levels of government, there are also non-governmental actors that play important roles in drinking water management, from source to tap, either directly or indirectly. The offloading of drinking water responsibilities from provincial and territorial governments to local governments poses an issue particularly in rural and remote places due to the capacity issues described above. This can lead to greater roles or reliance on third-party groups, including private water systems, watershed management committees, contractors and consultants, industry (e.g., oil and gas, forestry), and the general public (e.g., consumers and landowners) (Ramalho *et al.*, 2014). This increasing reliance on a diverse set of local-level actors, combined with context-specific drinking water challenges, make place-specific innovations in drinking water service delivery imperative, particularly for rural communities. The following case studies provide two examples.

Adoption of information and communication technologies: case study on the role of software in asset management activities in rural Newfoundland and Labrador, Canada

Asset management is a way to improve the lifespan and functioning of infrastructure (Bakker, 2007; Heare, 2007). In rural Newfoundland and Labrador, issues with adequate asset management practices have been identified. Best practices in asset management require knowing where your infrastructure is located and the condition of your infrastructure, having digital mapping, performing regular maintenance, and ensuring adequate human capacity to manage and maintain water systems. Previous research has found that asset management activities are lacking in rural Newfoundland and Labrador. The most common issues identified related to asset management were limited maps or as-builts[1] of water infrastructure, a lack of organized leak-detection programs, little evidence of strategic planning for moving forward with solutions, and an absence of sufficient funding to run the water system relative to the expenses associated with operating the system (Minnes and Vodden, 2017). Quite simply, it is very difficult to properly manage a drinking water system without knowing where it is located underneath the ground and its current condition, or without having adequate financial and human capacity to operate and maintain the system. Yet, this is the unfortunate reality in many remote and rural communities in Newfoundland and Labrador (Minnes and Vodden, 2017).

Electronic programs – such as ones that can help with inventory of infrastructure and help to assess appropriate water rates and financial planning for water systems – have been identified as a means to help small water system managers perform asset management (US EPA, 2003). These programs vary in cost and ease of use, and are even available through open source software. Open source programs lower costs for water managers or consultants helping communities with asset management, which in turn lowers the cost of the end product for users (Van den Berg, 2017). There are also pay-for-service electronic programs that can be used to assist with municipal asset management. One such program is TownSuite Mapping LITE, which is a municipal software package provided by PROCOM Data Services Inc (TownSuite Municipal Software, 2017). A case study was conducted in 2014 studying the use of this software in the rural community of the Town of Centreville-Wareham-

Trinity (population 1,147) (See Figure 9.1) (Statistics Canada, 2017). The town amalgamated the previously separate towns of Centreville, Wareham, and Trinity in 1992. The amalgamated town has two separate drinking water systems spread out amongst these formerly separate communities, making management of infrastructure more complex (Holisko *et al.*, 2014).

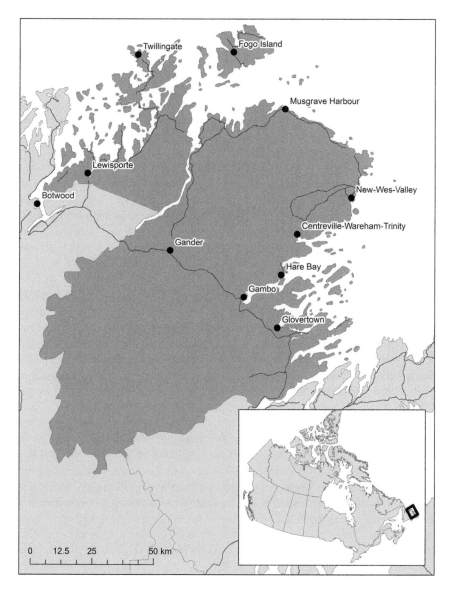

Figure 9.1 Map of the Town of Centreville-Wareham-Trinity.
Map credit: Myron King

In Newfoundland and Labrador, 40% of residents live in the St. John's census metropolitan area, with the majority of other residents living in smaller, more rural communities (Statistics Canada, 2017). For example, there are 529 public drinking water systems, with 68% of these systems serving 500 or fewer people, and 16% of these systems serving populations of 501–1,500. Therefore, the majority of public drinking water systems in Newfoundland and Labrador are in communities of 1,500 residents or fewer (Minnes and Vodden, 2017). As with many rural communities in Newfoundland and Labrador, the Town of Centreville-Wareham-Trinity was struggling with asset management and with as-builts and plan drawings not being available electronically, and sometimes being hand drawings that were out of date. Additionally, there were often multiple hard copies of the same document, resulting in inconsistencies as updates to the infrastructure were recorded on different versions. All together these issues made it very difficult for water operators and other town staff to perform asset management activities, including proper and timely response to emergency situations (e.g., leaks) (Daniels, 2014). The mayor of Centreville-Wareham-Trinity explained:

> Our primary concern was to digitize our valuable records, as a means of preserving them. Another key objective was to make it very easy for the maintenance staff, to help them identify where water lines are in the ground, rather than having to go through boxes of files, which could take hours on end, to look for [the] record of lines that had been buried in the ground over 30 years ago.
>
> (Daniels, 2014, pp. 1–2)

The TownSuite Mapping LITE software is specifically designed to address the issues that Centreville-Wareham-Trinity was having with managing the documentation of their water infrastructure. The software helps to digitize records and keep infrastructure records up to date, and assigns spatial identifications to assets for analysis and map making using geographic information systems (GIS). The program also has features to assist with capital budgeting to predict lifespans of infrastructure and when maintenance would be required (TownSuite Municipal Software, 2017). Since the full implementation of TownSuite Mapping LITE in Centreville-Wareham-Trinity, council, town staff, and water operators have noted the benefits of having digitized maps. For example, where curb stops can be located on an interactive map, town staff are now easily able to retrieve curb stop locations and other as-built drawings. This has also cut down on the time town staff spend looking for these documents. Furthermore, information specific to individual curb stops and as-builts can be emailed to water operators while they are in the field, saving additional staff time. Limited staff resources in many rural Newfoundland and Labrador communities increase the importance of such savings. The goals Centreville-Wareham-Trinity had of integration of data, preservation of important documents, and increased accessibility of records were achieved through the implementation of asset management software. Additionally, town staff are now able to better visualize water infrastructure in the ground (Daniels, 2014).

Small local governments often face challenges modernizing to new technologies and software due to limiting factors such as lacking the internal expertise to evaluate options and operate technologies, as well as limited tax bases and revenue sources to pay for such modernizations (Kot *et al.*, 2011; Minnes and Vodden, 2017). The cost of the TownSuite Mapping LITE program starts at $14,000 plus taxes. This cost includes the software, training, scanning, and setup. Additionally, there is an annual fee that includes software updates and technical support for the users at the municipality. Initial funding for TownSuite Mapping LITE was acquired by Centreville-Wareham-Trinity through federal Gas Tax[2] funding. There are financing options for those without the upfront funds (Daniels, 2014). Open source software provides lower cost options, with the coding and interface being available for free or at minimal cost. However, open source software can pose additional challenges for rural communities as it can require additional technological comfort and expertise. This makes the role of third-party expertise crucial. In the case of Centreville-Wareham-Trinity, they received, and continue to receive through an annual fee, a great deal of technical assistance from PROCOM Data Services Inc.

As with other types of rural service delivery, when geography allows, regional approaches to asset management can be a valuable option when considering the feasibility of introducing new technologies, with associated financial and human resource demands. For example, regional operators do not typically replace local operators, but they are highly qualified human resources who can oversee the operation and maintenance of water and wastewater systems. Regional operators in Newfoundland and Labrador have provided key capacity for asset management (Duffet, 2016). By successfully identifying several surrounding communities able to share or reduce the cost of staff (e.g., highly qualified operators) and equipment (e.g., asset management software), such regional operators can provide greater expertise in initiating at least simple versions of more sophisticated asset management electronic programs and performing related tasks. Regional operators could assist with data management activities such as digitizing maps and working with communities and local water operators to accurately locate infrastructure and plan for asset management activities. For example, a regional operator assisted in these activities in Centreville-Wareham-Trinity (Daniels, 2014). There is a need in Newfoundland and Labrador, however, for further incentives and sustained support for regional-level initiatives, such as staff (e.g., operators), service sharing, and drinking water management initiatives to enable their expansion (Minnes and Vodden, 2017).

A crucial part of sustainable drinking water management and proper asset management is having trained human resources. Issues with uncertified water operators have been identified in rural Newfoundland and Labrador. These issues include uncertified operators being less likely to perform asset management activities, as well as less likely to meet provincial requirements in checking chlorine residual in tap water (Minnes and Vodden, 2017). In addition to the potential of regional water operators noted above, the next section will speak to innovations in water operator training that are being made in rural British Columbia.

Rural appropriate operator training: case study on the Water Smart Peer-to-Peer Operator Training program in British Columbia, Canada

At a basic level, water operators are the people who operate and maintain drinking water and wastewater systems – ensuring the delivery of a critical service. 'Certified' operators have proven their credentials (i.e., education, experience, examination results) against a set of established criteria and standards to a constituted body (Environmental Operators Certification Program, 2014). Organizations like the Canadian Water and Waste Association (CWWA) and the Environmental Operators Certification Program (EOCP) believe that people operating the treatment and distribution of water and the collection and treatment of wastewater should be certified in order to minimize risk to human health and the natural environment, as well as to efficiently operate and care for infrastructure systems (Canadian Water and Wastewater Association, 2012b; Environmental Operators Certification Program, 2014). Certifying agencies across Canada develop their certification programs based on the standardized principles of the Association of Boards of Certification (ABC) guidelines – *Operator Certification Program Standards* (Canadian Water and Wastewater Association, 2012b). However, while the majority of certifying bodies across the country use the standardized Canadian exams produced by the ABC, not every province or territory in Canada requires certified operators. Only in Alberta, British Columbia, Nova Scotia, Ontario, and Saskatchewan is certification required (Canadian Water and Wastewater Association, 2012a).

As noted above, to gain certification requires proof of specific training and education, as well as on the job experience. Maintaining certification requires ongoing education for operators, or Continuing Education Units (CEUs). As with any professional body, continuing education ensures that these professionals have the up-to-date knowledge and skills needed to maintain a safe and effective system (Canadian Water and Wastewater Association, 2012b). There are recognized tangible benefits of certification, including the level of expertise of operators and standardization of practices, as well as other benefits such as networking among peers (Breen, 2016; Canadian Water and Wastewater Association, 2012b).

However, despite the recognized benefits, hiring and retaining certified water operators is not without challenges, particularly in rural and remote places. This technical/human capacity is a critical element of water systems, but one in which rural areas often lack sufficient capacity to undertake best practices in water management (Minnes and Vodden, 2017). Certification in particular comes with costs to the individual operators (e.g., education and annual association dues) and to their employers (e.g., wages of skilled professional). For employers, the cost of hiring and maintaining certified water operators not only includes wages, but the cost of training – including the cost of the courses, but also travel, accommodation, and replacing the operator while they are gone (Breen, 2016; McGowan, 2015; Speed, 2014). Additionally, certified operators are in high demand, with low recruitment and high attrition within the industry

(McGowan, 2015). This can create competition among employers to attract and retain certified operators.

Rural and remote communities are disproportionately disadvantaged when it comes to attracting and maintaining certified operators (Minnes and Vodden, 2017). For example, these communities are often unable to provide competitive professional salaries, which can lead to the poaching of operators by larger communities from smaller ones. Many small systems have only one or two certified operators, placing a great deal of responsibility on those working in these positions. If these individuals are not compensated accordingly, this can make recruitment and retention challenging. These issues are compounded by more typical issues of attraction in rural places (e.g., lack of spousal employment). As well, the distance from many rural and remote communities to certified training centres can increase the costs associated with maintaining certification, and thus of employing certified operators. Additionally, traditional training has been noted to be inappropriate for the rural and remote context – often focusing on systems and technologies that are not used in rural places. This calls into question the value for cost of such training, particularly when geographic proximity and availability, rather than the skills needed, guide decisions about what courses are attended (Breen, 2016; McGowan, 2015). Traditional training methods are also typically classroom learning, as opposed to providing hands-on experience (Breen, 2016).

In order to maintain certification, water operators in British Columbia must complete CEUs through the EOCP (Environmental Operators Certification Program, 2014). In the rural Columbia Basin region of southeast British Columbia, it was observed that the challenges discussed above were causing substantial hurdles for maintaining certification of operators, as well as to professional development and development of local capacity (McGowan, 2015). Within the region, challenges were identified with the traditional classroom-based model of training, including a lack of hands-on learning, cost, contextually inappropriate material, logistical challenges (e.g., temporary staff replacements), a lack of opportunities for training related to local interests (e.g., water loss management), and – perhaps most importantly – that training delivered by contractors or consultants limited the ability to develop sustainable, in-house capacity (McGowan, 2015). In order to ensure training meets the needs of communities, the CWWA suggests that local governments should work with educational institutions to develop and maintain educational programs that provide operators with appropriate skills and knowledge (Canadian Water and Wastewater Association, 2012b). However, the challenges described can also be addressed through alternative approaches. Collaborative approaches, particularly at a regional scale, have been noted as holding potential for capacity development (Minnes and Vodden, 2017). In an effort to address training-specific challenges, a collaborative effort was led by staff from the cities of Cranbrook and Castlegar, alongside Columbia Basin Trust's Water Smart program,[3] the EOCP, and four other communities within the region, to develop and test the infrastructure and process of supporting an official peer-to-peer (P2P) training approach to allow operators to earn CEUs through on-the-job mentorship.

In many ways, the Columbia Basin provided an ideal region for such a project. This mountainous region is home to a population of approximately 150,000 people, spread out over 80,000 km^2 (see Figure 9.2) (Columbia Basin Trust, 2017). The region is home to over 2,000 known drinking water systems, the vast

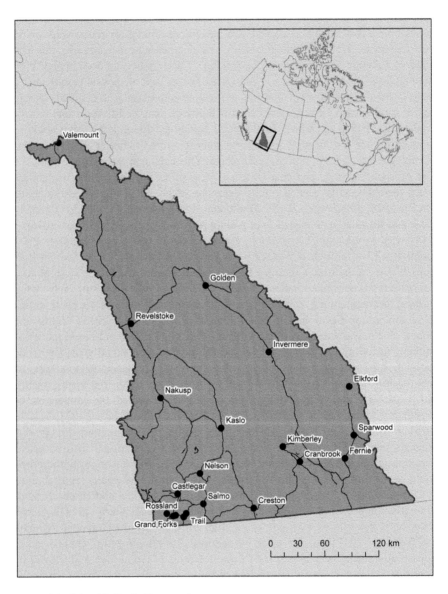

Figure 9.2 Columbia Basin Trust region.

Map credit: Aita Bezzola

majority of which are small systems with fewer than 500 connections (Breen, 2017). The drinking water systems within the region are characterized by those challenges described throughout this chapter, including capacity limitations (human and financial), aging infrastructure, and complex economic and development histories (Breen and Markey, 2015).

Prior to the collaborative pilot project, staff at the cities of Castlegar and Cranbrook had been working independently on P2P training, both in-house and with other, neighbouring, systems. By working together, alongside partner organizations, they were able to bring together the attention and resources needed to develop a formal program. The purpose of the P2P program was to "encourage and formalize the flow of knowledge between water operators while allowing operators to earn EOCP credits" (Breen and Hamstead, 2016, p. 62). The basic assumption behind the P2P program was that certified operators who had demonstrated particular expertise with a specific process or piece of equipment would be able to teach a short (half hour to three hour) course to their peers in a one-on-one or small-group setting. Each course followed a simple five-step process:

1. The trainer explains the role of the equipment and the specific component that is the topic of the training activity.
2. The trainer demonstrates the activity while describing each step.
3. The trainer coaches the trainee through the activity.
4. The trainee repeats the activity with minimal direction from the trainer.
5. The trainee describes the role of component, safety, and operational precautions to be taken; the process to be followed; and the tools necessary to complete the task.

An online training registry was developed to allow participating staff to register as trainers and/or trainees and to submit the necessary documentation (e.g., lesson plans). Once a course was delivered, both the trainer and trainee received CEUs, with the number of CEUs ranging depending on the course – lending both purpose and credibility to the process. Administrative support was provided to access and use the online platform. The Water Smart program staff acted as a driver by providing both support and a third party to whom communities were responsible. The flexibility within the online system allowed each community to have its own 'page'. Trainers could choose to offer courses strictly at an in-house/internal scale, sub-regionally, or regionally. This allowed for trainers to gain both experience and confidence, while setting the stage for future regional collaboration.

An external review was conducted during the pilot program, involving 29 participants from six communities ranging in size from 400 to 20,000, and totaling a population of almost 52,000 (Statistics Canada, 2017). The pilot project resulted in five registered providers, eight registered instructors, 21 courses registered, 17 courses delivered, 48 operators reached, and 9.35 CEUs earned (Breen, 2016). The external review explored the effectiveness of the approach, interviewing managers, trainers, and trainees at the start and end of the pilot. The identified benefits of the P2P approach included the expected benefit of providing hands-on training that was considered directly relevant to local systems, as well as fulfilling the hopes for professional development, including the taught skills, but also communication and mentorship skills, as well as enhanced staff confidence (Breen and Hamstead, 2016; Breen and McGowan, 2017). It is important to note that all participants felt that P2P was not a replacement for traditional training, but a complement to it (Breen, 2016). It was observed that the small-group nature of the approach helped participants to overcome reluctance to ask questions and fear of failure in learning new skills – promoting better understanding of the tasks and the system (Breen, 2016). Particularly, there is the potential for the P2P approach to be applied when considering the adoption of new technologies, such as those discussed in the previous section.

It is important to note that the P2P program was not without challenges – chief among them being the need for both internal and external support (Breen and Hamstead, 2016). Internally, the support needed was administrative, as well as from managers and decision makers to afford staff members time to develop their courses (Breen, 2016). External support and encouragement, particularly that provided by Water Smart staff, was also seen as critical. Many participants expressed doubt that such a program could continue without the push from an external body such as Water Smart (Breen, 2016).

Despite the success of the pilot project, the P2P program has encountered a number of challenges since the end of the pilot, including loss of key champions at the community scale, the end of the Water Smart program, and changes to the online training registry that have made it more difficult to use, particularly when unsupported. However, the P2P idea and five-step process remain useful, and have been transferred to other sectors (e.g., the P2P approach is being used at a high-techmaker space within the region), suggesting the continued merit of the approach.

Conclusion

In this chapter, we have argued that, in order to improve the management and operations of rural drinking water systems, innovative, rural-friendly, place-based solutions are needed to tackle the infrastructure deficit and to increase the sustainability of rural drinking water service delivery. The first case study demonstrated the potential and opportunity for technology to assist with asset management, while the second focused on developing a contextually appropriate approach to building human capacity in order to attract and retain skilled professionals. While asset management and the development of human capacity are only two elements of sustainable drinking water systems, they are two areas where it is possible to take action for improvement quickly in the short term as well as into the future.

What do the two case study examples presented have in common? First, both demonstrate innovation, one in product and the other in process. In this case, innovation is seen in a broad sense, including things that are both new and new in their application within a particular locale (i.e., the technology was innovative for the location where it was being applied). In both cases, these innovations emerged out of necessity and out of defined needs. Second, the two examples demonstrate a place-based approach, wherein the ideas were tailored to be made appropriate to the local context and capacity. In being designed to be scalable, serviceable, and flexible to address the particular challenges of rural and remote communities, these examples are fundamentally different from traditional, status quo models. Third, both examples capitalize on the benefits of taking a collaborative, regional approach in facing challenges and overcoming barriers to sustainable rural drinking water systems, including human capacity, financial capacity, and large, spread-out geographies. These three elements lend themselves to the transferability of these two examples to other rural and remote places.

However, the examples detailed illustrate potential pitfalls to future success as well. Financial capacity, in particular the creation of sustainable fiscal frameworks, is critical (e.g., full-cost accounting, protecting natural infrastructure, having a reserve fund, and charging sufficient water rates). Without a minimum amount of financial capacity, it is simply too difficult to embark on any initiative that extends beyond basic system function, and even that may be unachievable. Both examples also demonstrate the issues that face rural and remote places in recruiting and retaining skilled human capacity. We also see the importance of third-party organizations and programs, including supporting grants or incentive programs, as well as the role that external organizations can play in setting expectations and holding communities accountable. Also recognized is the critical role of individual champions. Unfortunately, as demonstrated in the case of the P2P training approach, short-term, temporary programs can often be too short to substantially change a system or culture, and have the potential to cause unintentional damage when time, effort, and commitment are put into ideas that have support in concept and are successful in the short term, only to be lost before any real momentum is gained. This raises questions as to where the overarching responsibility for longer-term implementation lies. With drinking water, this responsibility rests largely with local government; however, their limitations in terms of capacity, as well as their level of other responsibilities, have been outlined above. Rural communities range in capacity, including the size and expertise of their public works departments – from those that have fully staffed departments to those that do not. Many rural communities, both in the case studies and across Canada, do not have engineers on staff or access to engineers other than through paid consultants. Furthermore, their public works staff includes either one full-time, part time, or volunteer employee with other responsibilities besides drinking water system management. This lack of adequate staffing related to water management further limits their technical and human capacity to adopt new technologies.

Rural and remote communities are increasingly held to the same standard of service delivery as urban centres. In the case of drinking water systems, one can argue that the accessibility of safe drinking water is something that should be equal across communities of all types. And yet communities differ in their circumstances, often resulting in rural communities struggling or failing to meet standards or expectations for service delivery. As the case studies provided suggest, what perhaps is the most pertinent lesson for rural service delivery is that opportunities for innovations that will assist in meeting these challenges exist. Flexibility, tailoring of approaches to suit particular places, and considering strategies for long-term implementation of such new approaches are also critical in pursuing sustainable rural drinking water services.

Notes

1 As-built drawings are the original design drawings of the infrastructure and should include any revisions of changes to the infrastructure (e.g., design changes, relocations, rerouting of distribution systems) (Ellis, 2015).
2 The Gas Tax Agreement was released in conjunction with the New Building Canada Plan to "provide predictable, long-term, stable funding for Canadian municipalities to help them build and revitalize their local public infrastructure while creating jobs and long-term prosperity" (Infrastructure Canada, 2014, p.1).
3 The Water Smart program was a collaborative, regional-scale program focused on conserving water and building local capacity. While highly successful, the Water Smart program was shuttered by the Columbia Basin Trust in 2016 (Columbia Basin Trust, 2016).

References

Bakker, K. (ed.). 2007. *Eau Canada*. Vancouver, BC: UBC Press.
Baldwin, J.R. and Dixon, J. 2008. *The Canadian productivity review infrastructure capital: What is it? Where is it? How much of it is there?* Ottawa, ON: Statistics Canada.
Breen, S.P. 2016. *Tracking and evaluation: Water smart peer-to-peer operator training. Water smart peer-to-peer pilot project: Final report*. Columbia Basin Trust. Available online at http://eocp.ca/wp-content/uploads/2016/04/Water-Smart-P2P-Study-Final-Report-March-29.pdf. Accessed May 9, 2018.
Breen, S.P. 2017. *From staples theory to new regionalism: Managing drinking water for regional resilience in rural British Columbia*. Thesis. Vancouver, BC: Simon Fraser University.
Breen, S.P. and Hamstead, M. 2016. "Tapping in-house expertise: Columbia Basin Water operators deliver peer-to-peer training leadership", *Watermark* 25(2): 62–63.
Breen, S.P. and Markey, S. 2015. "Unintentional influence: Exploring the relationship between rural regional development and drinking water systems in rural British Columbia, Canada", *The Journal of Rural and Community Development* 10(3): 52–78.
Breen, S.P., and McGowan, J. 2017. "Peer-to-peer training approach offers alternatives to traditional classroom training", *CWWA Water Source* Spring: 10–12.
British Columbia Water and Waste Association. 2014. *Position statement: Sustainability of small water systems*. Vancouver, BC: BC Water Waste Association. Available online at https://bcwwa.org/resourcelibrary/2014-Mar-11SustainabilityofSmallWaterSystems.pdf. Accessed May 9, 2018.
Burleton, D. and Caranci, B. 2004. *Mind the gap: Finding the money to upgrade Canada's aging public infrastructure*. TD Bank Financial Group. Available online at www.td.com/

document/PDF/economics/special/td-economics-special-infra04-exec.pdf.   Accessed May 9, 2018.

Canadian Council of Ministers of the Environment. 2004. *From source to tap: Guidance on the multi-barrier approach to safe drinking water federal-provincial-territorial committee on drinking water.* Winnipeg, MB: Federal-Provincial-Territorial Committee on Drinking Water; Canadian Council of Ministers of the Environment. Available online at www.ccme. ca/files/Resources/water/source_tap/mba_guidance_doc_e.pdf. Accessed May 9, 2018.

Canadian Water and Wastewater Association. 2012a. *Operator certification.* Available online at www.cwwa.ca/faqmunicipal_e.asp#certification. Accessed May 9, 2018.

Canadian Water and Wastewater Association. 2012b. *Policies.* Available online at www. cwwa.ca/policy_e.asp. Accessed May 9, 2018.

CBCL Limited. 2012. *Managing municipal infrastructure in a changing climate.* St. John's, NL: Government of Newfoundland and Labrador, Department of Environment and Conservation.

Coad, L. 2009. *Improving infrastructure management: Municipal investments in water and wastewater infrastructure.* Report November 2009. Canada: Conference Board of Canada. Available online at www.probeinternational.org/EVfiles/10-115_CanCompete_ WaterInfrastructure_WEB_2.pdf. Accessed May 9, 2018.

Columbia Basin Trust. 2016. *The Columbia Basin water smart initiative: Building sustainable futures for community water use.* Castlegar, BC: Columbia Basin Trust. Available online at https://ourtrust.org/wp-content/uploads/downloads/2016-12_WaterSmart_Sum mary_FINAL.pdf. Accessed May 9, 2018.

Columbia Basin Trust. 2017. *Our story.* Available online at http://ourtrust.org/about/our-story/. Accessed February 27, 2017.

Cook, C., Prystajecky, N., Ngueng Feze, I., Joly, Y., Dunn, G., Kirby, E., Özdemir, V., and Isaac-Renton, J. 2013. "A comparison of the regulatory frameworks governing microbial testing of drinking water in three Canadian provinces", *Canadian Water Resources Journal* 38(3): 185–195.

Daniels, J.K. 2014. *TownSuite mapping LITE (+Scanning) and managing municipal water systems: Spotlight on the town of Centreville-Wareham-Trinity.* Environmental Policy Insitute, Memorial University of Newfoundland-Grenfell Campus. Available online at http://nlwater.ruralresilience.ca/wp-content/uploads/2013/04/Case-Study_TownSuite-Mapping-LITE_FINAL.pdf. Accessed May 9, 2018.

Duffet, I. 2016. *Regional water and wastewater operator pilot program.* Presentation. Gander, NL: Newfoundland and Labrador, Department of Municipal Affairs and Environment. Available online at www.mae.gov.nl.ca/waterres/training/adww/2016/12_Ian Duffett.pdf. Accessed May 9, 2018.

Ellis, R. 2015. *As built drawings.* Chaska, MN: Questions and Solutions Engineering. Available online at www.qseng.com/publications/rte/01/esmag01jun.htm. Accessed December 4, 2017.

Environmental Operators Certification Program. 2014. *Program guide.* Burnaby, BC: Environmental Operators Certification Program. Available online at http://eocp.ca/wp-content/uploads/2016/01/EOCP-Program-Guide-19_01_2016.pdf. Accessed May 9, 2018.

Federation of Canadian Municipalities. 2012. *Canadian infrastructure report card. Volume 1: 2012 municipal roads and water systems.* Ottawa, ON: Canadian Construction Association; Canadian Public Works Association; Canadian Society for Civil Engineering; Federation of Canadian Municipalities. Available online at https://csce.ca/wp-content/ uploads/2012/06/Infrastructure_Report_Card_ENG_Final1.pdf. Available May 9, 2018.

Health Protection Branch. 2013. *Small water system guidebook.* Victoria, BC: Government of British Columbia, Health Protection Branch. Available online at www2.gov.bc.ca/assets/gov/environment/air-land-water/small-water-system-guidebook.pdf. Accessed May 9, 2018.

Heare, S. 2007. "epa communiqué : Achieving sustainable water infrastructure", *American Water Works Association* 99(4): 24–26.

Hill, C., Furlong, K., Bakker, K., and Cohen, A. 2008. "Harmonization versus subsidiary in water governance: A review of water governance and legislation in the Canadian provinces and territories", *Canadian Water Resources Journal* 33(4): 315–332.

Holisko, S., Speed, D., Vodden, K., Sarkar, A., and Moss, S. 2014. *Developing a community-based monitoring program for drinking water supplies in the Indian Bay Watershed: A baseline study of surface water quality, contamination sources and resident practices and perceptions.* Report. The Harris Centre – RBC Water Research Outreach Fund, 2012–2013. Available online at www.mun.ca/harriscentre/reports/arf/2012/12-13-DWARF-Final-Vodden.pdf. Accessed May 9, 2018.

Infrastructure Canada. 2014. *The federal gas tax fund: Permanent and predictable funding for municipalities.* Available online at www.infrastructure.gc.ca/plan/gtf-fte-eng.html. Accessed May 9, 2018.

Kennedy, E., Roseland, M., Markey, S., and Connelly, S. 2008. *Canada's looming infrastructure crisis and gas tax agreements: Are strategic connections being made?* Working Paper # 2. Burnaby, BC: Simon Fraser University.

Kot, M. Castleden, H., and Gagnon, G. 2011. "Unintended consequences of regulating drinking water in rural Canadian communities: Examples from Atlantic Canada", *Health and Place* 17(5): 1030–1037.

Locke, W. 2011. *Municipal fiscal sustainability: Alternative funding arrangements to promote fiscal sustainability of Newfoundland and Labrador municipalities - the role of income and sales tax.* St. John's, NL: Municipalities Newfoundland and Labrador.

Maxwell, S. (ed.). 2008. *The business of water: A concise overview of challenges and opportunities in the water market (a compilation of recent articles from journal AWWA).* Denver, CO: American Water Works Association.

McGowan, J. 2015. *Concept paper for peer-to-peer training.* Unpublished. Cranbrook, BC.

Ministry of Community and Rural Development Local Government Infrastructure and Finance Division. 2010. *Local government infrastructure planning grant program: Program guide.* Victoria, BC: Ministry of Community Services, Local Government Infrastructure and Finance Division.

Minnes, S. and Vodden, K. 2017. "The capacity gap: Understanding impediments to sustainable drinking water systems in rural Newfoundland and Labrador", *Canadian Water Resources Journal/Revue canadienne des ressources hydriques* 1784(March): 1–16. DOI: 10.1080/07011784.2016.1256232.

Mirza, S. 2007. *Danger ahead: The coming collapse of Canada's municipal infrastructure.* Report. Ottawa, ON: Federation of Canadian Municipalities. Available online at www.mississauga.ca/file/COM/FCM_infrastructure_report.pdf. Accessed May 9, 2018.

Pollalis, S., Georgoulias, A., Ramos, S., and Schodek, D. 2012. *Infrastructure sustainability and design.* New York: Routledge.

Ramalho, C., Will, A., Macleod, J., and van Zyll de Jong, M. 2014. *Exploring the sustainability of drinking water systems in Newfoundland and Labrador: A scoping document.* Corner Brook, NL: The Harris Centre RBC Water Research and Outreach Fund. Available online at http://nlwater.ruralresilience.ca/wp-content/uploads/2013/04/FINAL_Rural-Drinking-Water-Scoping-Document_June11_Submitted-to-HC.pdf. Accessed May 9, 2018.

Robinson, J., Berkhout, T., Burch, S., Davis, E., Dusyk, N., and Shaw, A. 2008. *Infrastructure and communities: The path to sustainable communities*. With S. Sheppard and J. Tansey. Victoria, BC: Pacific Institute for Climate Solutions.

Santora, M. and Wilson, R. 2008. "Resilient and sustainable water infrastructure", *Journal-American Water Works Association* 100(12): 40–42.

Speed, D. 2014. *Community administrators survey results*. Corner Brook, NL: Environmental Policy Insitute, Memorial University of Newfoundland-Grenfell Campus. Available online at http://nlwater.ruralresilience.ca/wp-content/uploads/2013/04/Water-Administrator-Write-Up_FINAL_FINAL.pdf. Accessed May 9, 2018.

Statistics Canada. 2017. *Census profile*. 2016 Census. Statistics Canada. Available online at www.12.statcan.gc.ca/census-recensement/2016/dp-pd/prof/index.cfm?Lang=E. Accessed December 4, 2017.

TownSuite Municipal Software. 2017. *TownSuite asset management*. Available online at https://townsuite.com/asset-management. Accessed December 4, 2017.

UNESCO International Hydrological Programme. 2014. *Water in the post-2015 development agenda and sustainable development goals*. Discussion Paper. Paris, France: United Nations Educational, Scientific, and Cultural Organization; International Hydrological Programme. Available online at http://unesdoc.unesco.org/images/0022/002281/228120e.pdf. May 9, 2018.

US EPA. 2003. *Asset management: A handbook for small water systems*. United States Water. Available online at https://nepis.epa.gov/Exe/ZyPDF.cgi/2000261D.PDF?Dockey=2000261D.PDF. Accessed May 9, 2018.

Van den Berg, T. 2017. *2017 CRRF opening keynote panel: Putting small communities on the map again*. Canadian Rural Revitalization Foundation Conference. Nelson, BC: Rural Policy Learning Commons. Available online at www.youtube.com/watch?v=nK4b07xfflQandfeature=youtu.be.

10 Remotely connected?

A comparative analysis of the development of rural broadband infrastructure and initiatives

Wayne Kelly and Michael Hynes

Introduction

The popular and commercial development of the internet over the recent past has been astonishing and has initiated and enabled comprehensive changes in virtually every facet of human activity, leading to the creation of a 'network society' (Castells, 2002). It has enabled new forms and systems of communication; supported new sources and stores of information; facilitated the development of new businesses and new kinds of media; and allowed new forms of political, social, and cultural expression to emerge. New information and communication technologies (ICTs), improved connectivity, and sustainable infrastructure can be used to help make rural and remote communities more resilient to future challenges and shocks, and generate more equity between rural and urban populations. However, realizing new technology's universal utility will require investment that increases access to ICTs in remote, often low-productivity, areas, and the development of innovative services and applications that cater to the particular needs of rural and isolated communities. Research has found that internet connectivity benefits these areas by helping to overcome geographical isolation; promote access to resources, services, and opportunities; and encourage better social interactions and community attachment, which lowers the possibilities of outward-migration and stimulates economic development (Whitacre *et al.*, 2014a). The growth and prevalence of high-speed broadband allows greater flexibility in working hours and location, for instance, and the low cost and instantaneous sharing of ideas, knowledge, and skills have made collaborative work easier (Stone *et al.*, 2017). It allows workers to remotely access other computers and information stores easily from any access point on the network. In addition, many people now use the Web to access news, weather, and sports; to plan and book holidays; and to pursue their own personal interests. The importance of an 'information society' for maintaining and strengthening human rights has been argued (Klang and Murray, 2005; WSIS, 2003, United Nations, 2016), and although this position has been challenged (Cerf, 2012; Skepys, 2012), the internet can play a crucial role in promoting civil and political engagement (Feezell *et al.*, 2016; Kent and Zeitner, 2003; Smith *et al.*, 2009). The digital economy is defined as the new economic paradigm, which is marked by an increasing reliance on ICTs and digital technologies (Hadziristic, 2017), providing a wide range of economic and social opportunities and benefits. For communities and businesses, regardless of location, this means that they need to adopt digital technologies in order to successfully participate in that economy (Whitacre *et al.*, 2014b) as, for instance, the ability to complete transactions online is now an essential part of running a successful business in the 21st century (Kuttner, 2016).

Many governments recognize the role digital technologies now play in enabling economic activity and social development and have set out broad, ambitious plans to promote investment in internet access, such as the *Digital Agenda in Europe*, to capitalize on these opportunities (Deloitte, 2014). This chapter looks at the overall plans and strategies for the development of rural broadband infrastructure in both Canada and Ireland. It examines the current digital divide between urban and rural regions, provides an indication of pressures and barriers to the roll-out of fixed telecommunications infrastructure to rural communities and economies, and offers two separate case studies, from both Canada and Ireland, of rural broadband provision and initiatives in pursuit of the best way forward for the delivery of service to rural, remote, and isolated regions. These case studies will look at two separate models and mechanisms for delivering high-quality rural broadband when market forces are reluctant or failing to do so. Given the geographical, cultural, and social differences between rural and remote communities in Canada and Ireland, such case study investigations will better inform our overall understanding of the success, or otherwise, of innovative ways of delivering fixed broadband services to rural and isolated communities globally. They, furthermore, will serve to improve policy design with regards to the delivery of broadband services to these areas, helping to develop more resilient and sustainable rural communities and economies.

Rural broadband digital divide

While fixed broadband can provide rural communities with significant social and economic opportunities, these regions are often at a geographical disadvantage in accessing such essential service. Rural and remote communities typically do not have the same quality access to broadband and digital technology infrastructure, at comparable costs, as their urban neighbours (Townsend *et al.*, 2015). This gap is referred to as the 'digital divide' in both academia (Broadbent and Papadopoulos, 2012; Gallardo *et al.*, 2018; Roberts *et al.*, 2017; Townsend *et al.*, 2013; Warren, 2007) and public policy (CRTC, 2016; FCM, 2014; Government of the UK, 2017; OECD, 2001, 2011). Unfortunately, as the internet and digital technologies evolve, research shows that the digital infrastructure divide for rural communities persists; cell phone reception and fibre access are both examples of technologies improving faster than rural areas are able to keep pace (Salemink *et al.*, 2015). Indeed, rural areas often face the challenge of a 'double digital divide'; on the supply side, they lag behind in terms of the provision of next-generation access (NGA) infrastructure, and on the demand side, many rural communities lack the basic skills and knowledge to exploit the potential of digital technologies in terms of social and business development (ENRD, 2018). This is especially worrying and challenging as research has consistently suggested rural communities could benefit most from internet access (Ashton and Girard, 2013; Roberts *et al.*, 2017). One of the internet's uppermost contributions is removing geographic limitations to communication, knowledge access, and service delivery, as well as enabling and supporting economic transactions (Hallstrom, 2017).

Similar to other basic services and utilities, delivering broadband to rural areas is a challenge based on location, distance, and population density (Salemink *et al.*, 2015; Townsend *et al.*, 2013). These challenges frequently result in market failure in delivering rural fixed broadband. The persistence of monopoly-type structures in the provision of broadband infrastructure, the costs of implementation, and the lack of sufficient customer base results in little financial incentive for private-sector operators (Ashton and Girard, 2013; Kelly *et al.*, 2009). With market failure occurring across some rural communities in North America (Hupka, 2014; Rajabiun and Middleton, 2014), the UK (Ashmore *et al.*, 2015; Townsend *et al.*, 2015), and Ireland (Carnegie UK Trust and the Plunkett Foundation, 2012), other strategies and options are needed to ensure that efficient and effective broadband service is made available to such communities. Moreover, rural communities are unique due to cultural, demographic, and economic realities. As a result, one single broadband solution does not fit all rural communities (McNally *et al.*, 2017). In light of this diversity and ongoing market failure, rural broadband scholars assert that multiple methods of broadband delivery are needed to ensure that this basic but essential service is delivered to everyone, regardless of their location and population size (Kelly and McCullough, 2016; Pant and Hambly Odame, 2016; Philip *et al.*, 2017; Roberts *et al.*, 2015; Salemink *et al.*, 2015). One such alternative for delivering rural broadband is community broadband initiatives (Ashmore *et al.*, 2016; Kelly and McCullough, 2016). Community broadband initiatives are local broadband projects that "may involve focusing on one area of digital support or stimulating demand within a community to attract better services from the private sector or they may include engaging in local authority-led plans" (Ashmore *et al.*, 2016, p. 4). There are numerous examples of community broadband initiatives in Canada, the UK,[1] and Ireland,[2] while the United States has more than 200 municipal broadband initiatives delivering high-quality broadband in lieu of relying solely on market forces.[3] Another critical aspect of promoting wider availability and adoption of broadband is ensuring that an affordable fibre backbone infrastructure is in place. Sharing existing and new fixed infrastructure may be one strategy for achieving this goal more quickly and efficiently than simply letting the market take its course.

Rural broadband: the Canadian perspective

At the start of the 21st century, Canada was a leader in connectivity, ranked by the Organisation for Economic Cooperation and Development (OECD) as one of the top countries globally in internet access (McNally *et al.*, 2017). However, Canada has not been able to maintain that position, especially in rural broadband access. Reliance on market forces and a lack of national leadership slowed the pace of developing connectivity in rural Canada and has meant that other countries outperform Canada in internet access (FCM, 2014; International Telecommunications Union, 2017; OECD, 2017). Unfortunately, while Canada continues to recognize the importance and needs of rural regions in regard to broadband access, successive federal governments have continued to rely on market delivery (Rajabiun and Middleton, 2014). Relying on market forces, even with subsidies, to deliver broadband to all rural regions is not realistic and the geographic and demographic make-up of much of rural Canada makes it very challenging and unappealing for private-sector service providers to offer affordable and suitable internet service that is competitive with urban and global connections (Gallardo, 2016; Townsend *et al.*, 2015). In areas not being effectively served by the private sector, rural Canada requires alternative broadband delivery options if the private-sector options are reluctant or failing to deliver competitive broadband (Kelly and McCullough, 2016; Pant and Hambly Odame, 2016).

Another challenge in Canada is that the federal government shelved the broadband plans and strategy put forward by the National Broadband Taskforce (NBTF) in 2001, which would have continued to prioritize investment in broadband and identified the need for an emphasis on rural broadband access and on alternative non-market-based delivery models (NBTF, 2001). Over the next decade, while other countries adopted national broadband plans, Canada hesitated and failed to establish a broadband strategy (OECD, 2011). Whereas other countries set universal broadband access targets of 35 Mbps, Canada's target remained at 5 Mbps, only recently changing in 2016. As a result, Canada dropped from 2nd to 12th in internet access and continues to fall behind other OECD countries (McNally *et al.*, 2017). Change is underway, however, as the Canadian Radio-television Telecommunications Commission (CRTC) recently established a new minimum standard for broadband connectivity. This new standard was presented in December 2016 and consisted of 50 Mbps download speeds, 10 Mbps upload speeds, and unlimited data. The speeds included in this standard are intended to be a minimum or a start point for Canada (CRTC, 2016).

CRTC state of rural broadband in Canada information

CRTC's new basic standard for internet access of 50 Mbps download, 10 Mbps upload, and unlimited data returns the spotlight to Canada's urban-rural digital divide. Examining this divide in CRTC's 2017 monitoring report, it is revealed that when you include the upload speeds and unlimited data, only seven out of Canada's ten provinces and three territories offer that quality of internet access in rural areas (CRTC, 2017). To highlight the market failure to deliver competitive internet infrastructure options to meet CRTC's basic standards of 50 Mbps in this regard, only 41% of rural households have access to such broadband availability (CRTC, 2017). By comparison, 100% of urban centres with populations exceeding 29,000 are able to meet this standard.

In regard to the demand side or adoption of the internet, subscription to internet services meeting CRTC's basic standards is growing rapidly in Canada, with 26% of Canadians subscribing to 50-Mbps internet connections – a five-fold increase from 2013 when that rate was only 5% (CRTC, 2017). More Canadians are subscribing to faster internet services. However, due to a lack of availability in rural areas, these new subscriptions are predominantly in urban locations (*ibid.*). Overall, Canada's CRTC reports illustrate that, while internet connectivity is improving, rural internet is not keeping pace in either access or adoption with urban regions, and the digital divide continues to grow.

Canadian case study

The Canadian case study examines one region's efforts to bridge the urban-rural digital divide with a community broadband initiative that is a partnership between three rural municipalities and the school division they are located in. It will specifically focus on some of the activities and outcomes in one of the municipalities – Hamiota. However, the partnership with the other two munici-palities, Yellowhead and Prairie View, and Park West School Division, provides essential context and background. This regional and community-based broadband initiative is located in western Manitoba, in a rural region more than 100 kilometres from the nearest city. Hamiota has a population of 1,225, and the other two municipalities, Yellowhead and Prairie View, have populations of 1,948 and 2,088 respectively; the Park West School Division consists of 15 schools across six rural municipalities, including the three Fibre Co-op partnering municipalities.

Together these three rural municipalities and the school division formed Park West Fibre Co-op Inc. with the goal of bringing fibre optics communication options to the participating communities and to connect all schools in the division to fibre backbone (CBC News, 2017). This unique partnership is intended to help the school division and communities future-proof the region and ensure that the necessary infrastructure is in place to meet the needs and demands of businesses and individuals. The partners in this community broadband initiative have been working together since 2009. After approaching provincial and federal govern-ments to help improve local connectivity did not lead to solutions or support, the group decided to form their own cooperative and move forward on their own in 2016. The school division is utilizing fibre backbone to connect every school while the communities will roll out internet service to community members with much higher speeds but competitive pricing. More than 300 kilometres in fibre backbone were completed by mid-2017. Figure 10.1 illustrates where the Park West Fibre Co-op has laid fibre; the larger map from Kelly and McCullough (2016) also shows that the region served by the co-op was characterized by sparse population, which makes the provision of market-driven broadband a challenge.

As the CAO of Hamiota Municipality and the secretary treasurer of the fibre co-op noted in March 2018, fibre subscribers in Hamiota are receiving speeds between 600–900 Mbps, more than 100 times faster than the internet connection most residences had prior to the fibre option (Froese, 2018). Hamiota is using the funds collected from resident subscriptions to cover construction and

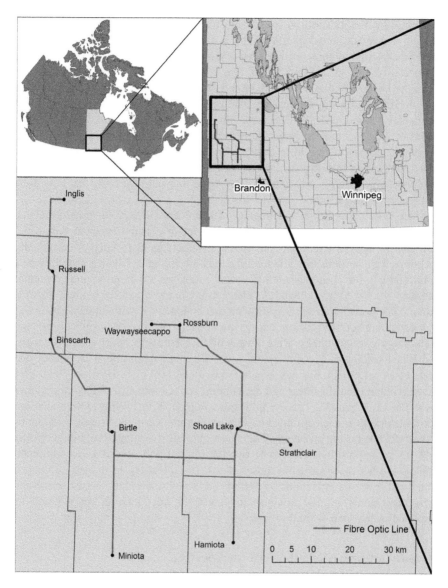

Figure 10.1 Southwest Manitoba Park West Fibre Optics Co-op Ltd. fibre backbone map
Map credit: Aita Bezzola, Adapted from www.pwsd.ca/News/PRESS-RELEASE-Apr-2016.pdf.

maintenance costs. To gauge community interest, address concerns, and increase uptake rates, the municipality of Hamiota has regularly engaged residents via town meetings and information materials. Hamiota began these meetings in November 2016 to discuss opportunities and concerns with community members and businesses. Additional meetings followed over the next few months, as the planning for the fibre roll-out continued. In May 2017, a FAQ session was held to help answer technical questions and a 'fibre to the home' demonstration was held in the high school once it was connected in June 2017 to showcase the connectivity and speeds and to increase interest in subscribing to the service. Hamiota began installation of FTTH and fast wireless to the rural residences in the municipality in the summer of 2017, and the switch to go live with the fast connections was flipped in December 2018. These meetings and demonstrations were an essential tool to help address concerns and confusion about both the benefits and costs of the fibre to the premise service. The community held two more community meetings in January and March 2018 to update the community on connection progress and to address any questions and concerns that new users of the service might have.

Since installation began, community stakeholders engaged in research related to broadband use have stated that there is a lot of demand for both bandwidth and data in the municipality. Community stakeholders also stated at community meetings that, in these first few months, they are trying to balance installing new connections with troubleshooting recent hook-ups and providing customer service – with the high demand for connectivity being a positive but real learning curve.[4] The potential of high-quality internet connectivity without data limits was a focal point for the discussions as participants inventoried applications and technology uses that they were aware of in urban and other connected rural regions. Participants are anticipating that the fibre connections in homes and businesses will help businesses and individuals alike in Hamiota, and they identified numerous business and agricultural applications that will be undertaken once fibre is available. These applications ranged from hosting online auctions for livestock to implementing precision agriculture via radio-frequency identification (RFID) tags in livestock. Participants are also very interested in accessing cost-saving applications like voice over IP (digital phone service over internet) and cutting television or cable subscriptions. Local leaders, businesses, organizations, and residents in Hamiota are all excited to use their high-speed internet and proud that some of the province and country's best internet service will be provided in their small rural region.

Broadband provision: the Irish perspective

Ireland has long supported the development of the country as a digital society, and since 2004, there have been four government initiatives to expand and improve broadband infrastructure, all of which have helped to some degree. But significant problems remain. While broadband has got faster, and more places than ever are served, a significant geographical region of the country still lacks commercial coverage. With basic service now available throughout Ireland, the focus has shifted to the delivery of high-speed fixed broadband to rural and remote areas of the country. Ireland has one of the highest proportions of people living in non-urban sites among EU states (European Commission, 2017), and concerns for rural areas and rural development maintain considerable priority on the policy agendas of both Ireland and the EU.[5] Such focus has changed from an early preoccupation with agriculture to maintaining viable and sustainable rural communities. Reduced dependency on farming and other primary industries is paralleled by local economies being more readily identified with new service industries, increased counter-urbanization, and associated commuting patterns; all have worked to transform Irish rural society over the recent past (McDonagh, 2007). A unique feature of the Irish rural countryside is the prevalence of one-off housing developments. A typical one-off house built in the last 20 years is frequently a bungalow of concrete block construction, leading to environmental journalist Frank McDonald coining the term 'bungalow blitz' in a series of articles in the 1980s (McDonald, 2018). Proponents of such development argue that it helps slow down rural depopulation, and that people should have the right to build on land they own. But the practice has been criticized for the environmental and aesthetic burden it imposes and the high cost of providing services to such disparate development. Nevertheless, rural and low-density regions in Ireland are often underserved, and broadband coverage in these locations remains considerably lower than national coverage across EU member states (European Commission, 2016). Satellite broadband remains the most pervasive technology in terms of overall coverage, while digital subscriber line (DSL) continues to be the most widespread fixed access technology. Very-high-bit-rate digital subscriber line (VDSL) remains the preferred NGA technology, with coverage increasing in the 12-month period to mid-2016, making it the fastest-growing fixed broadband technology. VDSL continues to be the key driver of NGA coverage growth across the EU, the European Commission maintains.

The National Broadband Plan (NBP) was published in August 2012 and provides a commitment to the provision of NGA broadband services to all Irish citizens and businesses, regardless of location (Department of Communications Energy and Natural Resources, 2012a).[6] The plan was expected to take between three to five years to complete and involved a state subsidy of up to €600 million. The plan aimed to radically change the broadband landscape in Ireland through a combination of commercial and state-led investment, and all parts of Ireland would have access to a modern and reliable high-speed broadband network capable of supporting current and future generations. The National Broadband Scheme (NBS) was co-funded by the European Regional Development Fund (ERDF) and was designed to deliver basic, affordable broadband to target regions across the country in which services were inadequate (Department of Communications, Climate Action, and Environment, 2017a). Following a competitive tendering process, the department awarded the contract to the telecommunications operator 'Three', who rolled out the scheme over 22 months, finishing in October 2010.[7] In early 2017, the largest telecommunications operator in Ireland, 'Eir', made a commitment to government to connect 300,000 homes in rural Ireland with fibre broadband on a commercial basis (Kennedy, 2017), but the sale of the company in December 2017 to a French telecoms investor put these commitments under pressure (Irish Examiner, 2017). Overall, broadband coverage in Ireland improved during the 12-month period to mid-2016 but remains below the EU average on both national and rural-level coverage (European Union, 2016). This report maintained that NGA coverage still reached only 50.3% of rural homes, following a 5.4% increase in the 12 months to the end of June 2016.

A commitment to providing a minimum of 30 Mbps for every home and business in Ireland, no matter how rural or remote, was contained in the NBP (Department of Communications, Energy, and Natural Resources, 2012a). However, in a recent worldwide broadband speed league report, with figures compiled from a year-long study (Cable.co.uk, 2017), Irish internet speeds were said to be among the slowest in Europe.[8] Ireland fell behind 25 other European states, 21 of which are in the EU, with Estonia, Jersey, Slovakia, and Slovenia among the many countries and regions providing better broadband speeds. Analysis of over 63 million broadband speed tests worldwide revealed Ireland is 36th internationally, with an average speed of just 13.92 Mbps. Nonetheless, there has been some progress in regard to targeted roll-out projects, in particular the claim that the over 780 post-primary schools in Ireland now have access to 100 Mbps high-speed broadband (Department of Communications, Climate Action, and Environment, 2017b).

The lack of rural broadband connectivity has condemned Ireland to one of Europe's lowest rates of broadband uptake. The EU's latest official digital score-board, which compares how countries are doing against one another, found that 7% of rural homes in Ireland still do not have access to even basic broadband, and the uptake of fixed broadband remains well below the EU average (European Commission, 2017). More than half of the population still does not possess basic digital skills – below the EU average and even further behind the leading digital countries in the EU, the report states. A lack of adequate fixed telecommunication infrastructure, combined with underdeveloped computer skills, is resulting in an increasing double digital divide and social disparities between urban and rural areas in Ireland. Moreover, low population density, demographic patterns, and geographical challenges have resulted in reluctance of service providers to invest in certain regions, particularly the mid-west. Not only is market failure evident in this region, but it is unlikely there will be sufficient competitive pressure in the market to improve fibre-based backhaul services in the short to medium terms (European Commission, 2013). However, the normal espoused rule that the development of a technology should be left solely to the marketplace does not apply in the case of broadband, which promises an array of social and economic benefits ranging from distance learning to telemedicine, public safety, and democracy (Weiser and Firestone, 2008).

The Ireland case study

There have been some interesting initiatives in regard to the provision of fixed broadband backbone services to rural areas in Ireland over the recent past. The majority of the cost of laying fibre is in the physical works and not in the fibre itself. This emphasizes the advantages of sharing passive existing infra-structure such as telephone poles, train tracks, or road networks. It also means that huge redundancy or overcapacity can be, and usually is, built in with minimal cost through adding additional fibres. This reinforces the case that it makes little sense for multiple competing optical fibre 'highways' to be built in parallel (O'Siochrú, 2008). In Ireland, one example of this type of single network using existing passive infrastructure is the telecommunications duct installation on 95 kilometres of roadway on the M7/M8 corridor between Dublin and Cork, due to be completed by mid-2017 (Department of Communications, Climate Action, and Environment, 2016a); although, this completion date was not met (Department of Rural and Community Development, 2017). Another example is the Galway-Mayo Telecoms Duct (see Figure 10.2).

In February 2005, the Irish government approved the construction of 132 kilometres of telecommunications ducting alongside the gas pipeline that had been under construction from Athenry in County Galway to the Shell gas terminal at Bellanaboy, County Mayo.[9] In May 2007, the construction of an additional 24 kilometres of duct was approved from the main trunk at Castlebar, County Mayo, to the outskirts of Westport in County Mayo. The basis for this decision was that the building of the gas pipeline afforded a unique opportunity to build a new underground telecommunications backhaul infrastructure at marginal cost, which could provide the opportunity for backhaul connectivity to rural parts of County Galway and County Mayo, capable of supporting advanced fibre optic setup. The public monies paid out in 2005 and 2007 went to Bord Gáis Éireann (BGÉ), a commercial state-owned company, to construct the telecommunications duct network along the route of the gas pipeline. The duct was to be owned by the Department of Communications, Energy, and Natural Resources, who gave the go-ahead to BGÉ to build the telecommunications duct in February 2005 by way of a commitment to cover all reasonable costs. However, no explicit contract was put in place at the time.

Upon completion, in a public consultation document from March 2012, the department issued a public tender invitation to interested parties to tender for the completion of the telecommunications network for the provision and bringing into service of suitable connection points along the infrastructure to allow for telecommunications operators to connect to this important backbone infrastructure (Department of Communications, Energy, and Natural Resources, 2012b). Gas Networks Ireland (GNI), formerly Bord Gáis Éireann, had agreed to obtain access rights from landowners along the route to lay, install, and operate the infrastructure, and to make these available to the minister, pursuant to the terms of a license agreement. However, in late 2015 it became apparent that there were outstanding issues in relation to access rights obtained by GNI, which needed to be resolved before the fibre could be formally transferred to the minister. Officials have been working with the Chief State Solicitor's Office and the Attorney General's Office on measures to resolve this issue but, to date, this has not been concluded.[10]

Discussion

The National Broadband Plan in Ireland and Canada's CRTC have both established the importance of universal service for broadband and have identified it as

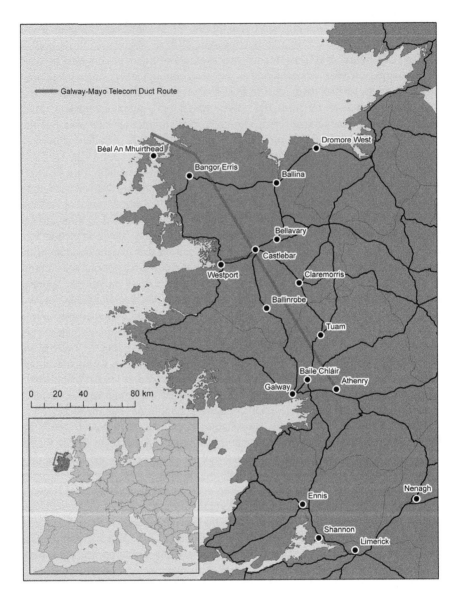

Figure 10.2 Galway-Mayo telecom duct route
Map credit: Aita Bezzola

a critical infrastructure for both economic development and quality of life in rural and remote regions. Although the less costly option of wireless backbone is increasing in bandwidth capacity, the preference is for optical fibre in terms of future-proofing the volume of traffic handled and the range of services carried.

Rolling out that backbone and allowing access to it in appropriate ways are necessary conditions for building a range of possibilities for using digital technologies and addressing the urban-rural digital divide. However, both countries have relied on using subsidies to enable and incentivize the market delivery of fixed broadband to rural regions, despite mounting evidence and research that market delivery alone is not successful long-term. Indeed, not all benefits of high-speed broadband can be monetized, and passive broadband infrastructure must be seen as a public good often meeting natural monopoly conditions outside urban areas (European Commission, 2015). The current policy positions and contexts in Canada and Ireland will be examined in this section against several key questions, identifying current challenges and potential opportunities.

Who is (or should be) leading the push for rural broadband?

In Canada, the basic standard of internet identified by the CRTC is not available in more than half of the country's provinces and territories. As a result, alternative models are needed and are increasingly being used in rural Canada, such as the community broadband initiative in Western Manitoba. The Municipality of Hamiota-Park West Fibre Optic Co-op provides an example of how fibre connectivity can be successfully delivered in rural areas. This project was implemented through a partnership between a school division and several of its communities. Research is increasingly calling for governments to support alternative models for broadband in rural regions, recognizing that market failure is delaying or preventing access to broadband, an essential infrastructure in today's global society. Federal and provincial governments need to provide increased financial support and capacity to help more rural regions take advantage of alternative models for broadband delivery.

In Ireland, the overall responsibility for the nationwide roll-out of fixed high-speed broadband rests centrally with the Department of Communications, Climate Action, and Environment. The emphasis has been largely to depend on private telecommunication industry investment and expertise, supplemented by government financial support. Notwithstanding some initial success in connecting rural communities, a commitment to deliver high-speed broadband access to 300,000 rural premises by the end of 2018 by commercial operator Eir may not be met, with 66.3% of homes within the target area remaining to be connected as of the third quarter of 2017 (Department of Communications, Climate Action, and Environment, 2017c). Further concerns with a commitment and application to tender to complete the overall National Broadband Plan were raised when Eir was taken over by a French telecoms tycoon at the end of December 2017, and in late January 2018, the company withdrew from the process, citing concerns over commercial, regulatory, and governance issues (Burke-Kennedy, 2018).[11] Overreliance on the market to roll out essential broadband services, it can be argued, is leading to increasing delays and will put the NBP under further pressure, specifically disadvantaging rural and remote regions of the country.

Meeting the targets established by the CRTC in Canada and by the government in Ireland will require alternative delivery models and increased collaboration with other stakeholders and levels of government in both countries. Governments need to recognize the need to engage all stakeholders and to support alternative delivery models in regions where the market is failing to provide quality infrastructure and service. Indeed, there is a strong case for direct government intervention in the event of market failure in those geographical locations not served by commercial operators (McDonnell, 2013). While seeking expertise and support from commercial telecommunications entities is both prudent and worthy, an overreliance on these private, market-driven organizations to provide fixed broadband services to commercially unviable, remote areas may be foolhardy. If access to broadband is to fulfil its objective as a social and economic good, then significant investment and control over its infrastructural and service development is needed by either local or central governments.

The role of local government

As reflected in Canada, local government needs to play an active role in expanding and securing broadband infrastructure. The case study in Canada illustrates that through creative partnership, in this case with their school division, small rural municipalities can deliver top-quality broadband access to residents and businesses. It requires commitment, creativity, and perseverance by local leaders, and trust and support from residents and businesses. There are increasing numbers of examples in Canada of small rural municipalities delivering world-class broadband in areas ignored or underserved by market forces. In Ireland, local government has a limited to non-existent role to play in the delivery of fixed broadband to rural and remote regions of the country. In comparison to other EU states, Ireland has a weak system of local government, largely as a result of strict central controls, the absence of financial independence, and a narrow functional range (Daemen and Schaap, 2000).[12] Moreover, Ireland is unique in the developed world in not having carried out any profound reform of its local government structure in the last 100 years. The sparse diffusion of functions and powers to local authorities is one of the distinguishing features of the Irish administrative system (Coakley and Gallagher, 2010). Thus, local government in Ireland lacks the utility required to compel or take serious action to address Ireland's rural and remote fixed broadband structural defect.

The way forward: a model for rural broadband delivery

Community broadband initiatives offer an alternative to market-force delivery of broadband infrastructure in rural regions, but they, too, face barriers. While they have the potential to deliver rural internet access, it is essential to understand the scale and challenge involved in delivering that access. These locally driven initiatives take substantial upfront financial commitment from communities (McNally *et al.*, 2017), while the capacity and leadership required at the local level to undertake a community broadband initiative is significant (Ashmore *et al.*, 2016). These challenges are not trivial and, combined with the unique nature of remote regions, it is not realistic to expect every rural region to undertake a local broadband initiative. Minimum requirements to replicate community broadband initiatives across rural regions are necessary, including government financial assistance and training supports, as well as local leadership and stakeholder collaboration. While this model provides an alternative approach for delivering broadband, most rural communities will not be able or willing to undertake it on their own. To complement the market approach for internet access, governments must consider similar subsidies and stimuli for community-broadband initiatives to those being applied to incentivize market delivery. Furthermore, utilizing existing road, rail, and other such fixed structural networks and configurations to provide the backbone necessary for the delivery of broadband to rural and remote areas is both practical and prudent.[13] However, regulation over the development and administration of this vital backbone infrastructure must remain steadfastly within the realm of government control, either local or central, in order for the potential of rapidly developing broadband services to be fully realized.

Conclusions

High-speed internet access is the new and evolving prerequisite for modern living and working, and high-quality fixed service broadband is something citizens demand and want, irrespective of where they choose to live. Broadband infrastructure is a transformative infrastructure, with the capacity to change what people do and the way they do it, and it can be particularly useful in promoting regional and rural development and reducing, if not eradicating, the limitations of living in more peripheral and isolated locations. But while states and regions espouse the development and roll-out of efficacious broadband infrastructure and services to all citizens, regardless of their location, pressures and barriers exist to accomplishing this in a timely and efficient manner. In the cases of Canada and Ireland, there is an overreliance on commercial market forces to provide this essential service on a judicious basis which, to date, has not always been productive. It is preferable to invest in long-term legacy infrastructure rather than short-term service delivery, and the case for more extensive roll-out of fibre is greater if considerations like the wider economic and social benefits to rural and remote areas are included, and the appropriate long-term evaluation period chosen. There are examples of government policies in other countries supporting the delivery of next-generation broadband in less populous areas and where there is market failure, and these should be used to inform the roll-out of the fixed broadband to remote areas of Canada and Ireland. However, the true indication of success will be reflected in the response and uptake of the service provided; therefore, cognizance of the unique makeup and support of communities needs to be better considered. In addition, the use of new and existing fixed infrastructural networks will play a central role in the provision of the backbone needed to provide an operational service. More consultation with local communities is necessary, as is more research into the unique circumstances that exist for each discrete rural or remote community, in order for broadband to truly live up to its potential as a transformative infrastructure and service in both Canada and Ireland.

Appendix

Glossary of acronyms

BGÉ   Bord Gáis Éireann
CRTC   Canadian Radio-television Telecommunications Commission
DSL   Digital subscriber line
ERDF   European Regional Development Fund
EU   European Union
FTTH   Fibre to the home
GNI   Gas Networks Ireland
ICT   Information and communication technology
ISP   Internet service provider
Mbps   Megabits per second
NBP   National Broadband Plan (Ireland)
NBS   National Broadband Scheme (Ireland)
NBTF   National Broadband Taskforce (Canada)
NGA   Next-generation access
OECD   Organisation for Economic Cooperation and Development
RFID   Radio-frequency identification
VDSL   Very-high-bit-rate subscriber line

Notes

1 Examples of community broadband initiatives in Canada and the UK include (1) O-Net, Olds, Alberta, Canada, (2) SMART, Parkland, Alberta, Canada, (3) Cybermoor Ltd., Cumbria, England, (4) Lothian Broadband, East Lothian, Scotland, and (5) B4RN, Lancashire, England.
2 The Ludgate Hub, Skibbereen, Co. Cork, is a good example of such an initiative. With the backing of significant local private investment, the hub aims to facilitate up to 75 people in a creative co-working environment with a long-term objective to create 500 direct jobs and 1,000 indirect jobs (see www.ludgate.ie).
3 See the Community Broadband Networks Website at muninetworks.org/content/muni cipal-ftth-networks.
4 Wayne Kelly is co-leading this research at Brandon University as part of the Examining Broadband with Rural Communities research at the Rural Development Institute. More information can be found here: www.brandonu.ca/rdi/files/2016/11/ RDI-Newsletter-Fall-Nov17-Final.pdf.
5 The CSO defines the rural population as those living outside settlements of 1,500 people (see www.cso.ie/en/releasesandpublications/ep/p-cp1hii/cp1hii/bgn/), while the Commission for the Economic Development of Rural Areas (CEDRA) defined rural as those areas outside the administrative boundaries of the five main cities of Dublin, Cork, Limerick, Galway, and Waterford (see www.chg.gov.ie/rural/rural-development/ cedra/).
6 The NBP, with a cost of up to one billion euros, originally sought to serve more than 900,000 premises and more than 1.8 million citizens on the wrong side of the digital divide.
7 Under EU State Aid rules this intervention was for a limited duration. The NBS ended following a 68-month operational period at midnight on August 25th, 2014, and

'Three' continues to provide a limited broadband coverage throughout NBS areas on a commercial basis.

8 Despite being among the slowest for average broadband speeds, Ireland was also found to be the eighth most expensive country in Europe for the average cost of broadband.

9 The Corrib gas project entailed the extraction of natural gas deposits off the northwest coast of Ireland. The project included the development of the Corrib gas field, construction of the natural gas pipeline, and a gas-processing plant on shore. The Corrib project faced strong opposition, with persistent protests against the development for close to a decade. Locals and activists were concerned about the safety of the project and its environmental impacts (Siggins, 2010).

10 The Telecommunications Services (Ducting and Cables) Bill, a bill developed to give access to the Galway-Mayo ducting and cables telecommunications network, passed the Seanad [upper house] second-stage reading in February 2018 and will now go forward for debate and approval in the Dáil [lower house] (Houses of the Oireachtas, 2018).

11 Last year, the other significant bidder, Siro - a joint venture between Vodafone and the Electricity Supply Board (ESB) - pulled out of the process, stating that it was unable to make a business case for continued participation. Concerns over the only remaining bidder, ENET, surfaced after the decision of Conal Henry to step down from the helm of that organization (Kennedy, 2018).

12 Local government is based exclusively on the counties created in the middle ages, too big for effective municipal government and too small to have a regional remit, and has little relevance to the modern structure of the Irish economy and population.

13 Some positive moves towards adopting this particular strategy are now evident. A 2014 EU directive formally set out approaches for greater access to civil infrastructure in order to provide the necessary incentives to telecoms operators to invest in future network infrastructure (EC, 2014). The primary purpose of the Broadband Cost Reduction Regulations in Ireland is to provide a framework of rights and obligations aimed at facilitating and reducing the cost of deploying high-speed public communications networks (Department of Communications, Energy, and Natural Resources, 2016b).

References

Ashmore, F.H., Farrington, J.H., and Skerratt, S. 2015. "Superfast broadband and rural community resilience: Examining the rural need for speed", *Scottish Geographical Journal* 131(3–4): 265–278.

Ashmore, F.H., Farrington, J.H., and Skerratt, S. 2016. "Community-led broadband in rural digital infrastructure development: Implications for resilience", *Journal of Rural Studies* 54: 408–425.

Ashton, W. and Girard, R. 2013. "Reducing the digital divide in Manitoba", *Journal of Rural and Community Development* 8(2): 62–78.

Broadbent, R. and Papadopoulos, T. 2012. "Getting wired@collingwood: An ICT project underpinned by action research", *Community Development Journal* 47(2): 248–265.

Burke-Kennedy, E. 2018. "Eir warned Naughten of 'deep flaws' before quitting broadband scheme", *The Irish Time* February 17, 2018.

Cable.co.uk. 2017. *Speed test data shows UK's average broadband speed is 16.51 Mbps.* Available online at www.cable.co.uk/news/new-broadband-league-shows-uks-average-speed-is-less-than-half–700001889/. Accessed December 12, 2017.

Carnegie UK Trust and the Plunkett Foundation. 2012. Rural broadband – reframing the debate. Available online at www.carnegieuktrust.org.uk/getattachment/49d067b8-5836-4906-bfe0-70978c6ca5e9/Rural-Broadband—Reframing-the-Debate.aspx. Accessed November 27, 2017.

Castells, M. 2002. *The internet galaxy: Reflections on the internet, business, and society.* Oxford: Oxford University Press.

CBC (Canadian Broadcast Corporation). 2017. *3 western Manitoba municipalities, school division form their own fibre optic internet service.* August 5, 2017. Available online at www.cbc.ca/news/canada/manitoba/rural-fibre-optic-internet-cooperative-1.4236694. Accessed December 2, 2017.

Cerf, V.G. 2012. "Internet access is not a human right", *New York Times*, January 4, 2012.

Coakley, J. and Gallagher, M. 2010. *Politics in the republic of Ireland.* 5th edition. New York: Routledge.

CRTC (Canadian Radio-television and Telecommunications Commission). 2016. Closing the broadband gap initiative. Available online at www.crtc.gc.ca/eng/Internet/Internet. htm. Accessed September 1, 2017.

CRTC (Canadian Radio-television and Telecommunications Commission). 2017. *Communications monitoring report 2017.* Available online at http://crtc.gc.ca/eng/publications/reports/PolicyMonitoring/2016/cmri.htm. Accessed September 1, 2017.

Daemen, H. and Schaap, L. 2000. Ireland: Associated democracy. In H. Daemen and L. Schaap (eds.), *Citizens and city: Development in fifteen local democracies in Europe*, (pp. 57–74). Rotterdam: Centre for Local Democracy.

Deloitte. 2014. *Value of connectivity economic and social benefits of expanding internet access.* London: Deloitte.

Department of Communications Energy and Natural Resources. 2012a. *Delivering a connected society a national broadband plan for Ireland.* Dublin: Department of Communications, Energy and Natural Resources.

Department of Communications Energy and Natural Resources. 2012b. *Public consultation on the bringing into use of a telecommunications duct infrastructure constructed alongside the gas pipeline between Ballymoneen Co. Galway and Bellanaboy County Mayo with a spur into Castlebar and Westport.* Dublin: Department of Communications, Energy and Natural Resources.

Department of Communications, Climate Action, and Environment. 2016a. *Report of the mobile phone and broadband taskforce.* Dublin: Department of Communications, Energy and Natural Resources.

Department of Communications, Climate Action, and Environment. 2016b. European Union (Reduction of cost of deploying high-speed public communications networks) regulations 2016. Available online at www.dccae.gov.ie/en-ie/communications/legislation/Pages/European-Union-(Reduction-of-Cost-of-Deploying-High-Speed-Public-Communications-Networks)-Regulations-2016.aspx. Accessed January 10, 2018.

Department of Communications, Climate Action, and Environment. 2017a. *National Broadband Scheme (NBS).* Available online at www.dccae.gov.ie/en-ie/communications/topics/Broadband/closed-schemes/national-broadband-scheme/Pages/National-Broadband-Scheme.aspx. Accessed January 20, 2018.

Department of Communications, Climate Action, and Environment. 2017b. *Schools 100 Mbps project.* Available online at www.dccae.gov.ie/en-ie/communications/topics/Broadband/pages/schools-100mbps-project.aspx. Accessed January 10, 2018.

Department of Communications, Climate Action, and Environment. 2017c. *Rural deployment progress.* Available online at www.dccae.gov.ie/en-ie/communications/topics/Broadband/national-broadband-plan/commercial-investment/Pages/Rural-Deployment-Progress.aspx. Accessed December 12, 2017.

Department of Rural and Community Development. 2017. *Action plan for rural development – First progress report.* Dublin: Department of Rural and Community Development.

ENRD. 2018. Workshop 1 – rural digital hubs. How to ensure that they successfully help rural businesses seize the opportunities of digitisation? European Network for Rural Development (ENRD) Seminar on 'Revitalising Rural through Business Innovation'. Brussels, March 2017. Available online at https://enrd.ec.europa.eu/sites/enrd/files/s4_rural-busi nesses-factsheet_digital-hubs.pdf. Accessed March 20, 2018.

European Commission. 2013. Next Generation (backhaul) Network (NGN) alongside a gas pipeline in Galway and Mayo. Available online at http://ec.europa.eu/competition/state_aid/ cases/243213/243213_1504550_221_2.pdf. Accessed October 23, 2017.

European Commission. 2014. *Directive 2014/61/EU of the European parliament and the council of 15 May 2014 on measures to reduce the cost of deploying high-speed electronic communications networks.* Brussels: European Commission. Available online at https:// ec.europa.eu/digital-single-market/en/news/directive-201461eu-european-parliament-and-council. Accessed January 2, 2018.

European Commission. 2015. *Note on the socio-economic benefits of high-speed broadband. B5/FM/IO.* Brussels: European Commission. Available online at http://europedirectpuglia. eu/files/Socio-economic-benefits-of-High-Speed-Broadband.pdf. Accessed December 2, 2017.

European Commission. 2016. *Broadband coverage in Europe 2016: Mapping progress towards the coverage objectives of the digital agenda.* Luxembourg: European Commission, Directorate-General of Communications Networks, Content and Technology.

European Commission. 2017. *Europe's digital progress report 2017.* Available online at http:// ec.europa.eu/eurostat/statistics-explained/index.php/Statistics_on_rural_areas_in_the_EU. Accessed February 10, 2018.

FCM (Federation of Canadian Municipalities). 2014. *Broadband access in rural Canada: The role of connectivity in building vibrant communities.* Available online at https://fcm.ca/ Documents/reports/FCM/Broadband_Access_in_Rural_Canada_The_role_of_connectivi ty_in_building_vibrant_communities_EN.pdf. Accessed December 28, 2017.

Feezell, J.T., Conroy, M. and Guerrero, M. 2016. "Internet use and political participation: Engaging citizenship norms through online activities", *Journal of Information Technology and Politics* 13(2): 95–107.

Froese, I. (2018). Rural internet co-op plugs away to connect customers. *Brandon Sun*, March 23, 2018. Available online at www.brandonsun.com/local/rural-internet-co-op-plugs-away-to-connect-customers-477705443.html.

Gallardo, R. 2016. *Responsive countryside: The digital age and rural communities.* Mississippi State, MS: Mississippi State University Extension.

Gallardo, R., Whitacre, B., and Grant, A. 2018. *Broadband's impact, a brief literature review.* Purdue University. Research and Policy Insights, Center for Regional Development.

Government of the United Kingdom. 2017. *Broadband delivery UK.* Available online at www.gov.uk/guidance/broadband-delivery-uk#history. Accessed November 27, 2017.

Hadziristic, T. 2017. *The state of digital literacy in Canada: A literature review.* Brookfield Institute. Available online at http://brookfieldinstitute.ca/wp-content/uploads/2017/04/ BrookfieldInstitute_State-of-Digital-Literacy-in-Canada_Literature_WorkingPaper.pdf. Accessed November 12, 2017.

Hallstrom, L. 2017. *Beyond infrastructure: Strategies to support adoption and realize benefits of broadband in rural Canada.* ACSRC Report Series #49-17. University of Alberta, Alberta Centre for Sustainable Rural Communities. ACRRS. Available online at www.cloudfront.ualberta.ca/-/media/augustana/research/acsrc/reports/acsrc/no-49-17. pdf. Accessed August 15, 2018.

Houses of the Oireachtas. 2018. *Telecommunications services (ducting and cables) Bill 2018.* Available online at www.oireachtas.ie/en/bills/bill/2018/13/. Accessed March 30, 2018.

Hupka, Y. 2014. *Findings on the economic benefits of broadband expansion to rural and remote areas.* University of Minnesota, Centre for Urban and Regional Affairs. CAP Report #188. Available online at www.cura.umn.edu/publications/search. Accessed December 12, 2017.

International Telecommunications Union. 2017. *Measuring the information society report 2017.* Available online at www.itu.int:80/en/ITU-D/Statistics/Pages/publications/mis2017.aspx. Accessed December 18, 2017.

Irish Examiner 2017. "Spotlight on national broadband plan as Eir sold again in €3.5bn transaction", *The Irish Examiner* December 21, 2017.

Kelly, T., Mulas, V., Raja, S., Qiang, C.Z.W., and Williams, M. 2009. *What role should governments play in broadband development?* Paper prepared for info Dev/OECD workshop on *Policy Coherence in ICT for Development,* Paris, September 10–11, 2009.

Kelly, W. and McCullough, S. 2016. *Research brief: State of rural information and communication technologies in Manitoba.* Brandon, MB: Rural Development Institute.

Kennedy, J. 2017. "Eir strikes deal to connect 300,000 rural homes with fibre broadband", *Siliconrepublic.* Available online at www.siliconrepublic.com/comms/eir-government-broadband-deal. Accessed April 4, 2017.

Kennedy, J. 2018. "National broadband plan hits another speed bump", *Siliconrepublic.* Available online at www.siliconrepublic.com/comms/national-broadband-plan-conal-henry-enet. Accessed March 5, 2018.

Kent, J.M. and Zeitner, V. 2003. "Internet use and civic engagement: A longitudinal analysis", *Public Opinion Quarterly* 67: 311–334.

Klang M. and Murray, A. 2005. *Human rights in the digital age.* London: The GlassHouse Press.

Kuttner, H. 2016. *The economic impact of rural broadband.* Washington, DC: Hudson Institute.

McDonagh, J. 2007. Rural development. In B. Bartley and R. Kitchin (eds.), *Understanding contemporary Ireland,* (pp. 88–99). London: Pluto Press.

McDonald, F. 2018. "'Bungalow Blitz' another nail in the coffin for towns and villages", *The Irish Time,* Tuesday February 13, 2018.

McDonnell, T.A. 2013. *The economics of broadband in Ireland: Country endowments, telecommunications capital stock, and household adoption decisions.* Doctor of Philosophy Thesis. J. E.Cairnes School of Business and Economics, National University of Ireland Galway.

McNally, M.B., McMahon, R., Rathi, D., Pearce, H., Evaniew, J., and Prevatt, C. 2017. *Understanding community broadband: The Alberta broadband toolkit* (compressed version). Edmonton, Alberta: University of Alberta.

NBTF (National Broadband Task Force). 2001. *Final report.* Industry Canada. Available online at http://publications.gc.ca/collections/Collection/C2-574-2001E.pdf. Accessed October 12, 2017.

O'Siochrú, S. 2008. *Rural broadband backbone: A case study of different approaches and potential.* Available online at http://gb1.apc.org/es/system/files/APCProPoorKit_PolicyAndRegulation_CaseStudyRural_EN_0.pdf. Accessed September 10, 2017.

OECD. 2001. Understanding the digital divide. Organisation for Economic Co-Operation and Development. Available online at www.oecd.org/sti/1888451.pdf. Accessed November 12, 2017.

OECD. 2011. *National broadband plans*. OECD digital economy papers 181. Paris: OECD Publishing. Available online at http://dx.doi.org/10.1787/5kg9sr5fmqwd-en. Accessed December 3, 2017.

OECD. 2017. *Key issues for digital transformation in the G20*. Available online at www.oecd. org/g20/key-issues-for-digital-transformation-in-the-g20.pdf. Accessed October 2017.

Pant, L.P. and Hambly Odame, H. 2016. "Broadband for a sustainable digital future of rural communities: A reflexive interactive assessment", *Journal of Rural Studies* 54: 435–450.

Philip, L., Cottrill, C., Farrington, J., Williams, F., and Ashmore, F. 2017. "The digital divide: Patterns, policy and scenarios for connecting the 'final few' in rural communities across great Britain", *Journal of Rural Studies* 54: 386–398.

Rajabiun, R. and Middleton, C. 2014. "Rural broadband development in Canada's provinces: An overview of policy approaches", *Journal of Rural and Community Development* 8(2): 7–22.

Roberts, E., Anderson, B., Skerrat, S., and Farrington, J. 2017. "A review of the rural-digital policy agenda from a community resilience perspective", *Journal of Rural Studies* 54: 372–385.

Roberts, E., Farrington, J. and Skerratt, S. 2015. "Evaluating new digital technologies through a framework of resilience", *Scottish Geographical Journal* 131(3–4): 253–264.

Salemink, K., Strijker, D., and Bosworth, G. 2015. "Rural development in the digital age: A systematic literature review on unequal ICT availability, adoption, and use in rural areas", *Journal of Rural Studies* 54: 360–371.

Siggins, L. 2010. *Once upon a time in the west: The Corrib gas controversy*. London: Transworld Ireland.

Skepys, B. 2012. "Is there a human right to the internet?", *Journal of Politics and Law* 5: 15–29.

Smith, A.W., Schlozman, K.L., Verba, S., and Brady, H. 2009. *The internet and civic engagement*. Washington, DC: Pew Internet and American Life Project.

Stone, K., Dagnino, E., and Martínez, S.F. 2017. *Labour in the 21st century: Insights into a changing world of work*. Newcastle upon Tyne: Cambridge Scholars Publishing.

Townsend, L., Sathiaseelan, A., Fairhurst, G., and Wallace, C. 2013. "Enhanced broadband access as a solution to the social and economic problems of the rural digital divide", *Local Economy* 28(6): 580–595.

Townsend, L., Wallace, C., and Fairhurst, G. 2015. "'Stuck out here': The critical role of broadband for remote rural places", *Scottish Geographic Journal* 131(3–4): 171–180.

United Nations. 2016. *Promotion and protection of all human rights, civil, political, economic, social and cultural rights, including the right to development*. United Nations General Assembly, Human Rights Council. Available online at www.article19.org/data/ files/Internet_Statement_Adopted.pdf. Accessed November 7, 2017.

Warren, M. 2007. "The digital vicious cycle: Links between social disadvantage and digital exclusion in rural areas", *Telecommunications Policy* 31(6): 374–388.

Weiser, P.J. and Firestone, C.M. 2008. *A framework for a national broadband policy*. Washington, DC: Aspen Institute, Communications and Society Program.

Whitacre, B., Gallardo, R., and Strover, S. 2014a. "Broadband's contribution to economic growth in rural areas: Moving towards a causal relationship", *Telecommunications Policy* 38(11): 1011–1023.

Whitacre, B., Gallardo, R., and Strover, S. 2014b. "Does rural broadband impact jobs and income? Evidence from spatial and first-differenced regressions", *The Annals of Regional Science* 53(3): 649–670.

WSIS. 2003. *Declaration of principles. Building the information society: A global challenge in the new Millennium*. Geneva: World Summit on the Information Society.

11 Pursuing alternative infrastructure arrangements to strengthen service provision in British Columbia, Canada

Laura Ryser, Greg Halseth, and Sean Markey

Introduction

As with many rural and small town places within developed economies, rural British Columbia (BC) has been experiencing an ongoing transformation of its human services and associated infrastructure. Since the 1980s, public policies have increasingly called upon local agencies and providers to devise more integrated or shared service arrangements as a part of 'bottom-up' community development (Argent, 2011; Paagman *et al.*, 2015). These policies, however, are challenging the transformative capacity of these same rural organizations by demanding change while withholding adequate funding to support the activities needed to build and maintain partnerships, by not following through with supportive policies or the training and mentoring needed for local groups to assess and successfully launch new service arrangements, and by not delegating an appropriate level of authority to accompany new mandates. Small communities are working through these changes while at the same time being burdened with the challenges associated with aging and inadequate infrastructure – much of which was developed in the period of resource frontier expansion (1950s – 1970s).

In this chapter, we present the findings from a research study on service provision in rural BC. We detail 16 cases of local governments and service leader organizations experimenting with innovative responses to address the challenges of restructuring and aging infrastructure in order to maintain and expand rural service provision. Specifically, we focus upon the use of co-location initiatives as a vehicle to pool financial resources and expand infrastructure capable of supporting local government and community service initiatives. Following this introduction, the chapter provides a brief overview of rural and small town BC, a review of the literature on service co-location, an outline of our research methodology, core thematic findings, and a discussion. The conclusion draws together key lessons from the cases and reflects upon the implications of, and standards for, innovation in rural and small town service delivery.

British Columbia context

British Columbia covers a large territory (944,735 km^2) yet its population is relatively small (4.6 million, 2016 Census). Much of that population is concentrated in the southwest corner of the province in the metropolitan Vancouver-Victoria region (2.84 million, 2016 Census). Aside from some regional centres such as Kelowna, Kamloops, Prince George, and Nanaimo, the settlement pattern in the rest of the province is marked by small communities – often with populations between 5,000 and 10,000 people. The mountainous landscape limits transportation routes and options, and thus the rural and small town settlements are often near key transportation crossroads or a local natural resource opportunity. While economic diversification for small communities has been a mantra in the province for decades, many of these small places are still dependent on a single natural resource industry base (Markey *et al.*, 2012).

BC's post-World War II development focused almost exclusively on the exploitation of natural resources (primarily minerals, oil and natural gas, coal, wood products, and pulp and paper). This era of 'province building' coincided with the post-war Keynesian public policy framework. The cumulative result was an expansion of natural resource industries into the rural and northern 'frontier' regions of the province. To support industrial opportunity, the provincial government undertook massive public investments in critical infrastructure such as hydro-electric projects, as well as highways and railway lines into regions that just a decade before were still using fur trade transportation methods and routes. The provincial government also invested heavily in new towns – called 'instant towns' – that came into existence in wilderness settings in a matter of two to three years (Halseth and Sullivan, 2002). The problem embedded within many of these places was that their economies were single-industry dependent in both design and focus. In the decades to come this would prove a difficult legacy to overcome.

In terms of local services, rural and small town places received a high level of investment commensurate with what would be expected from a Keynesian public policy framework. Schools and hospitals were built, as were recreation and cultural infrastructure. Services were professionalized and expanded as the rural and small town population grew from the early 1950s to the end of the 1970s. Many small places had provincial and federal government offices delivering a wide range of services. The private sector also expanded its provision of services in order to attract and retain employees and their families, as opportunities were growing during this period of economic expansion.

Since the 1980s, the governance regime in BC has transformed from a Keynesian public policy framework into a more neoliberal public policy framework defined by government withdrawal, reduced state expenditures, and a reliance upon market-based metrics and providers. This macro policy transition is found in many developed economies (Halseth and Ryser, 2018). Its application in BC has witnessed the aggressive closure of both provincial and federal government offices and services in rural and small town locations in favour of 'centralization' into regional hubs (Halseth and Ryser, 2006; Sullivan *et al.*, 2014). This pattern of cuts, closures, and regionalization was then repeated among private sector services. The withdrawal of resources was exacerbated in both public and private sector services by the rise of information technologies during the late 20th and early 21st centuries – with the refrain that the services and information was still available 'on our website'. Of course, accessing these websites was difficult as internet and mobile phone networks were much slower to expand into low-density rural and small town markets.

These changes in policy approach and in provision of services have been problematic for several reasons. First, critical services have been reduced at the very time that communities must respond to increasingly dramatic and dynamic economic change (Manson *et al.*, 2016; Ryser *et al.*, 2014). Services provide access to expertise and leadership, support networks, and links to assistive agencies whose staff are familiar with circumstances 'on the ground', and they are especially needed when there is the stress and challenges surrounding processes of economic change or collapse. Second, in addition to economic change, these post-war 'boom' communities are experiencing equally dramatic demographic changes as the resource-industry workforce ages in place and youth have been leaving in search of educational or employment opportunities. This is a process that Hanlon and Halseth (2005) have termed 'resource frontier aging' and it places an increased level of need and expectation upon local services – many of which have transitioned into being volunteer supported or delivered (Skinner *et al.*, 2014). Economic change and demographic transition, however, are both challenging the capacity of volunteers in rural and small town places to meet the increased demands placed on them by policy and community change (Ryser and Halseth, 2014).

The closure and regionalization of services have been especially challenging for older people and those with limited incomes. Services to support those living in poverty must be available and delivered locally as, by definition, these people lack the financial resources to take the time and expense of travelling to distant service centres or even to purchase the computer and internet services needed to get access to online information or forms (Ryser and Halseth, 2017). Older residents are also limited in their mobility, and the costs as well as the difficulty of traveling long distances in winter weather over mountain roads to access services places them, and others, at risk (Ryser and Halseth, 2012).

To meet these challenges, rural and small town places, as well as the volunteers and service providers in those places, are experimenting with innovative solutions by re-bundling existing resources in creative ways. In the following section, we review the body of literature related to the coordination, integration, and co-location (via partnerships and other mechanisms) of remaining local service providers in order to maintain or expand service delivery.

Services literature: co-location of rural and small town services

Shared service and infrastructure arrangements are not new. Research suggests that they have been pursued in the public, private, and non-profit sectors in many OECD countries such as Australia, Canada, Germany, the Netherlands, New Zealand, Norway, Scotland, Sweden, the UK, and the US since the 1970s (Browne, 2011; Malone and Anderson, 2014; Minas, 2014; Rumball-Smith *et al.*, 2014; Scottish Government, 2007; Stein *et al.*, 2011). They are a popular response to service restructuring as they have the potential to provide a number of benefits. Despite this potential, the study and application of co-location initiatives remains underdeveloped in rural areas (Van Belle and Trusler, 2005).

In terms of generalized benefits, shared service arrangements may allow organizations to expand their networks and connect with a broader range of expertise, information, and resources (Bauch, 2001; Community Foundations of Canada, 2009). These networks may be mobilized to build more support and legitimacy for initiatives, stronger co-operative relationships, and strategic alliances (Beachy *et al.*, 2010). They may also assist with financial efficiencies and providing better economies of scale (KPMG, 2014; Walsh, 2008; Whitfield, 2007). This includes potential savings by sharing equipment, vehicles, and staff in IT, payroll, human resources, and reception (Lennie, 2010).

Coordinated and shared arrangements provide opportunities to consolidate and reduce redundant programs and adopt more efficient and streamlined processes for referrals, delivering services, information management, and decision-making (Corbett and Noyes, 2008; Dollery and Akimov, 2007; Grant *et al.*, 2007). In terms of operations, these shared arrangements may create capacity to respond to service needs that are complex and often beyond the capacity or mandate of any single organization or are outside their previous geographic scope of service provision (Graves and Marston, 2011).

There can also be benefits in terms of opportunities to pool human capital in order to strengthen organizational capacity. They may be able to share knowledge, skills, and information about new requirements, new technologies, and other innovations (Janssen and Joha, 2006). It may allow organizations to become more competitive in attracting and retaining staff, and applying more consistent human resource practices (Evans and Grantham, n.d.). In turn, shared service and infrastructure arrangements can nurture cooperation, organizational change, and organizational learning. Further, by scaling up their capacity, organizations can achieve a stronger voice and better position themselves to advocate and negotiate with government.

The complexity of the model presents challenges around the development of partnerships and the very long time that it may take to identify and build appropriate space. Even with significant investment of time and energy in partnership development, co-location initiatives can struggle with different levels of interest and participation among partners. Change can bring fear, and organizations which have long operated independently and in their own space or facility may be wary of change towards co-location even when it makes sense along all of the parameters noted above. There may be concerns, for example, with costs, lease or rental agreements, suitability of space, or privacy of clients, etc. (Conway *et al.*, 2011; Rural/Urban Cost-Sharing Task Force, n.d.). Concerns about sharing equipment, books, and operational and maintenance costs, as well as concerns about the roles, responsibilities, and job security of different unionized personnel can also be a challenge (Blue-Howells *et al.*, 2008; Grimsey and Lewis, 2002; Sloper, 2004).

Many of the benefits and challenges associated with co-location may be generalized to the specifics of the model, and efforts to deliver services in new and innovative ways. As noted above, however, there are gaps when viewing alternative service provision through a rural lens. Small town dynamics, which may exacerbate the impacts of restructuring, present context-specific challenges. They may also present opportunities – for example, the tighter social networks often present in rural communities and the extent to which these may ease collaboration. In the following section, we outline our methodology and study dynamics.

Case research: co-location in the resource periphery

Our research team conducted 51 interviews in 35 communities to examine the best practices and key issues that must be considered by rural stakeholders and senior governments who wish to strengthen smart service and infrastructure projects in BC (see Figure 11.1). The research explored multi-purpose or co-location initiatives, service co-operatives, and multi-service or one-stop shop organizations. There was a series of best practices and key messages that can inform policies and investments to support the renewal of small communities in order to better position them to respond to the challenges and opportunities associated with the new rural economy.

For this chapter, we draw upon 16 case studies of local governments and service leaders around BC that specifically explored co-location initiatives (see Table 11.1). We explore critical themes involving funding, governance, site selection and design, human resources, and equipment and technology that shape the planning, development, and operations of these complex pursuits. Many of these initiatives were driven by bottom-up processes in which community stakeholders took the lead to fund, plan, develop, and maintain these shared infrastructure arrangements. There were other co-location arrangements in which both community and senior government stakeholders were engaged in the planning, development, and management of such facilities. As we will discuss below, however, these mixed-led pursuits did not always result in community ownership of shared or multi-purpose facilities that could otherwise strengthen the resiliency of organizations in these places.

Funding

By their very definition as small communities, stakeholders and organizations operate with smaller economies of scale that equip them with fewer resources. Local governments often make significant contributions to co-location initiatives by matching federal infrastructure funds. Local governments, however, face a dilemma when applying for grants that require them to prove they have the ability to proceed if funding is approved. They must decide whether to ask the community to borrow money for the multi-purpose facility before applying for the grant or after the grant is secured. Some communities benefitted from amenity fees associated with large developments, property sales, surplus and reserve funds, and a strong industrial tax base. Other communities, however, had a more limited tax base that made it difficult to acquire their share (i.e., one-third of the total project funds) of matching funding within provincial and federal infrastructure programs. These financial challenges have become more difficult

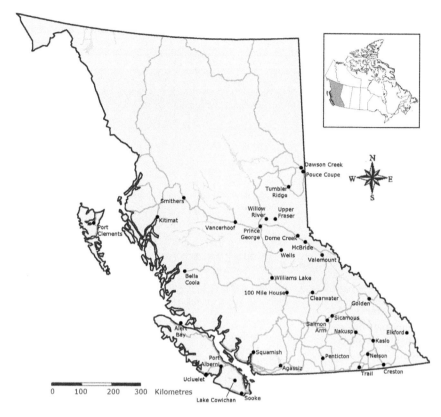

Figure 11.1 Locations of case studies in BC
Map credit: Kyle Kusch

as industries reduce community donation programs and, in a number of cases, reduce or close their operations.

Some communities pursued joint infrastructure initiatives with school districts in order to save rural schools (Charest and Chanter, 2011). Through the provincial Neighbourhood Learning Centre initiative, school districts became major financial contributors to multi-purpose facilities in smaller communities as they covered expenses for the construction of shared spaces, such as common hallways and reception areas, as well as signage, appliances, and equipment related to community spaces in these facilities. School districts also provided in-kind support by having their trades staff build shelving units or install equipment and appliances for shared community spaces in these facilities. There is no certain long-term way, however, to protect these rural assets from future school closure decisions. Rural stakeholders engaged with joint infrastructure partnerships with schools are continuing to build financial reserves in order to purchase multi-purpose facilities from school districts in the event of future closures.

Table 11.1 Matrix of co-located infrastructure arrangements driven by stakeholders

Initiative	Provincial-led	Mixed	Bottom-up	Operational	Build Assets/Reserves
Central Interior Community Services Co-op			•	•	•
Dome Creek Multi-Purpose Recreation Facility			•	•	•
East Line Activity Centre*	•			•	
Elkford Community Conference Centre			•	•	•
Fraser Valley Employment & Support Services Co-op			•	•	•
Gold Loom Co-location Project**			•		
Lake Cowichan Neighbourhood Learning Centre*, **	•				
Penticton and District Community Resources Society			•	•	•
Port Clements Multi-Purpose Building	•			•	•
Pouce Coupe Regional Multi-Purpose Centre*	•			•	
Sea to Sky Community Services			•	•	•
Sooke Co-op Assoc. of Service Agencies			•	•	•
Ucluelet Multi-Purpose Community Centre			•	•	•
Valemount Multi-Purpose Facility			•	•	•
Wells Multi-Purpose Facility*	•			•	•
Yellowhead Community Services	•			•	

* Note: Established through the Neighbourhood Learning Centre initiative.
** Note: Examples of shared infrastructure initiatives that were not successful in their development.

Small communities have received funding to support co-location or multi-purpose infrastructure programs through other provincial ministries, such as the Ministry of Community and Rural Development (Towns for Tomorrow[1]); the Ministry of Community, Sport, and Culture; and the Ministry of Jobs, Tourism, and Skills Training (StepUp BC/Labour Market Partnership). The constant restructuring of short-term provincial and federal funding programs, however, meant that rural stakeholders often did not have adequate time for communities to mobilize and engage in larger and more complex infrastructure projects. Opportunities for community planning and consultation were limited. Strict criteria for eligible expenses also impeded the early stages of developing joint infrastructure projects. Rural stakeholders struggled to justify the importance of paid coordinators to nurture relationships and planning for co-location initiatives. There were also complications to secure trust or foundation funding for co-location or shared infrastructure arrangements that are partly or wholly owned by school district partners due to concerns about investing funds to support senior government facilities, despite other uses contained in these assets.

Governance

In the early stages of development, the tendering and construction of co-location facilities were often managed by a school district or local government. although there were co-location initiatives managed by non-profits. After facilities were completed, these building committees ceased to exist. Local governments or community organizations assumed responsibilities for booking rental spaces and addressing complaints. Surprisingly, once the facilities transitioned into operations, formal management/tenant committees were rarely put in place (service co-operatives were the exception). Instead, tenants would raise issues with the local government or key partners in the facility's ownership as part of the conflict resolution protocols in lease agreements.

The governance and operations of these co-located or shared infrastructure arrangements have been guided by ownership and user agreements, risk and liability agreements, management agreements, conflict resolution protocols, health and safety plans, emergency protocols, maintenance agreements, and ad-hoc committees for specific topics. In some sites that were mixed or co-led with both senior government and local stakeholders in their development, final agreements for the ownership, use, and management of multi-purpose facilities created discomfort for non-profits that invested time and resources to raise funds, contribute expertise and, at times, manage site construction but were excluded from the security of ownership and control of building assets.

Site selection and design

In dispersed communities and rural regions, site selection was sometimes controversial due to the costs of extending physical infrastructure and proximity to different stakeholders. Once designated sites were in place, the design and functionality of facilities were addressed through common features such as shared reception areas; separate inside or outside access points; multi-purpose activity rooms, family rooms, and quiet activity rooms; shared whiteboard walls that could be folded away to create larger rooms; Wi-Fi, teleconferencing, videoconferencing, and audio-visual equipment; expanded storage space in the walls; shared staff rooms; storage for recreational equipment; banquet facilities; and common kitchen areas. Budget constraints meant that some co-location facilities needed to compromise certain design elements such as staff rooms, a first aid room, landscaping, air conditioning, divider walls, and the size of various rooms in the facility.

Different types of spaces required modifications to meet regulations and support activities. Wide observation windows, hand-washing sinks, private access, and sound-proofed confidential spaces were needed to support the work of community service and public health professionals. Ground floor access needed to be allocated to accommodate partners that provide services to seniors, young parents with strollers, or people with mobility issues. A number of small communities also used these projects to enhance energy efficiency through investments in passive solar heating systems for hot water, geothermal heating systems, infrastructure to collect water from rooftops for use in washroom facilities, and better insulation.

Human resources

Shared infrastructure arrangements have also transformed human resource strategies. Through co-location, there was greater stability and coverage for day-to-day operations. Shared administrative and financial staff supported communication and reporting with all levels of government, wrote grants, worked through procurement processes, processed local government applications for property tax breaks, submitted invoices, and completed income tax returns. Executive directors and staff were then able to spend more time delivering needed services. Human resources were expanded as a result of the savings acquired by sharing other staff.

After developing processes for secondment, organizations further shared program staff to provide coverage for those who were away, to deliver different or complementary program supports, and to deliver fundraising events. Co-located groups seized opportunities for joint recruitment and training on topics such as finance and administration, core training (i.e., first aid, occupational health and safety, conflict resolution, and cultural sensitivity training), employment services, or different types of therapy. There were some challenges working with different personalities and organizations which became protective and concerned about pooling their staff.

Stakeholders in co-located facilities were grappling with the intersection of different union workforces in their human resource strategies. These issues were complicated by the different sets of rules presented by different unions. Concerns emerged about the potential for one union to file grievances when the member of a different union assumes specific responsibilities in a consolidated or shared workforce. Ongoing conversations were taking place to resolve disputes between unionized and non-unionized stakeholders in co-located facilities that pursued shared human resources for tasks such as maintenance, reception, and administrative support.

Equipment, supplies, and technology

Co-location initiatives presented more opportunities to share resources to support day-to-day operations such as equipment, phone and internet service plans, multi-media infrastructure, and supplies, allowing service agencies to reduce costs and coordinate the use of shared spaces. The extent to which opportunities were taken to share resources amongst co-located organizations varied, however, and were underutilized or non-existent at times. Some organizations chose to remain self-contained. Despite a number of provincial and federal initiatives, there are still many small and remote communities around BC that do not have broadband infrastructure. The absence of technology infrastructure has limited the design and potential of multi-purpose facilities.

Discussion

As with any innovation, the test of its value is measured in terms of whether it benefits the users (or owners), or in this case communities. The threshold to maintain and continue to develop service innovation is complicated in rural and small town places. Rural communities have been undergoing complex and dramatic changes, primarily linked with cuts and withdrawals of key services but also associated with demographic change and new service demands. The threshold may be that the innovative response is simply better than nothing (i.e., co-location or service cut). This requires communities to either accept new baseline conditions of reduced service provision or call on volunteers to make up the difference. This may place additional strain on communities and weaken their overall community and economic development foundations.

Our cases illustrate that there are varieties of positives associated with co-location initiatives. Many of the examples serve as innovative responses to restructuring, responding either to cuts in funding; degraded and poorly maintained infrastructure; or new demographic, community, and economic development realities in their communities and regions. Community infrastructure gaps are being addressed. In terms of added value, first, we see that co-location has enhanced the stability of many operations. Consistency of service delivery represents a very important dynamic both in terms of being able to adequately serve rural populations and in communicating the stability of important community development foundations to internal and external audiences (i.e., it is important for the attraction and retention of both capital and labour). Second, co-location appears to be saving precious time for staff and volunteers – time that is being redirected into actual service delivery. Third, co-location appears to be helping to retain key services. Co-location is providing an adequate solution to the need for space, to funding challenges to maintain space, and to the sharing of resources to create a mutually supportive platform for a variety of services. Finally, co-location appears to enhance the flexibility and adaptability of service agencies to what are often rapidly changing conditions. Changes related to the shifting and often short-term priorities of governments, or to shifting economic conditions – for example the booms and busts associated with the resource economy – present tremendous challenges to maintaining services and/or having the resources to respond to peak times of critical need that are not captured in traditional funding formulas.

In terms of challenges, it is not uncommon for innovative responses to encounter resistance to change as a result of established practices and/or power relations, or bureaucratic, regulatory frameworks. Our case examples show evidence of this resistance; for example, challenges associated with navigating traditional funding requirements which are no longer compatible with collaborative arrangements. This is not to say that senior governments are on principle resistant to service innovation, but that bureaucratic processes of engaging have not caught up to the innovations taking place on the ground. Second, we see the cases struggling with new governance arrangements. Co-location and collaborative service delivery require new skills and new shared governance models. And finally, some of our cases illustrate the traditional problem associated with funding programs that support the establishment of hard infrastructure but not the necessary resources for actual operations. Non-profits are used to this catch-22. It is clear that program responses which are seeking to support smart service delivery, e.g. co-location, are not supporting the full spectrum of organizational operations; furthermore, there are programmatic barriers associated with jurisdictional confusion, or siloed government entities or ministries that are not being well coordinated surrounding programmatic spaces where their activities align or collide.

Conclusion

In this study, we examined 16 cases across rural and small town BC that are engaging with co-location initiatives as one way to support new or continued local service delivery. While such initiatives demonstrate commitment, hard work, and ingenuity, they remain isolated test cases of potential rural innovation. With a number of infrastructure programs across various provincial and federal ministries, there is no central hub for rural stakeholders to learn about different models and processes that have been used to develop co-location initiatives. There is limited understanding of ownership and user agreements, design features that can improve the functionality of multi-purpose spaces, risks and liabilities, and protocols to guide the development, operations, and maintenance of these facilities. Given the limited tax base of many small communities, greater flexibility is needed to support financing arrangements. Short-term funding programs do not provide adequate time for communities to mobilize and engage stakeholders in these complex projects. Such short-term funding programs can also interrupt the momentum for building relationships, planning, and mobilizing other assets for these initiatives.

More positively, co-location and other smart service delivery mechanisms open up tremendous potential and policy terrain for the revitalized co-construction of rural infrastructure. Co-construction refers to a new paradigm of governance whereby both top-down senior government actors and agencies and bottom-up communities and regions co-create territorially appropriate development initiatives (Bryant, 2010). This serves as an alternative to contextually blind top-down intervention, the capacity for which has ceased to exist following two decades of neoliberalism. It also provides an alternative to simple abandonment, recognizing that there is a role for senior governments to play while also demanding a greater level of engagement and responsibility for development processes from communities and regions themselves, their local governments, and service agencies. Co-location initiatives represent a functional example whereby co-constructed governance relationships may be built. It also presents an opportunity for senior government programming to establish policy and programmatic generalities that then may be shaped to suit local conditions.

Moving forward, greater political leadership is needed to designate ministries to lead supportive policies that can shape the implementation of new shared infrastructure and service arrangements (Farquhar *et al.*, 2006). Senior levels of government can also create shared services assessment teams or managers to provide advice and guide the development of these new infrastructure and service arrangements that are increasingly being requested through their policies (Lennie, 2010; Zeemering and Delabbio, 2013). Some provinces are already working to deploy teams capable of facilitating and providing advice for those pursuing shared infrastructure cost-sharing negotiations (Rural/Urban Cost-Sharing Task Force, n.d.). These changes will go a long way towards better positioning stakeholders in small communities to respond to the challenges and opportunities associated with rural change.

Based upon the stories of our interviewees, we are mindful of how challenging it is to create new ways of maintaining and operating services in rural areas. Any type of transition may be challenging, but the core issue is whether these operations are viable. Smart services need to be more than the last hope to maintain service delivery. It is clear from our research that co-location, as a form of smart service delivery, presents tremendous potential benefits. If it is viewed, however, as a way to cut core budgets while maintaining some form of degraded service presence, then it clearly is not an innovative response. Rather, it would serve as a façade of innovation, seeking to cover for more narrow objectives associated with austerity for austerity's sake. Real investment in rural services is still a necessary condition of innovation.

Note

1 Towns for Tomorrow was an infrastructure grant program launched in December 2006. Typically, provincial granting programs provide matching funds up to one-third of the project costs. Towns for Tomorrow, however, provided up to 80% of the funding needed for rural infrastructure projects. Small communities (under 5,000 people) cost-shared with the Province on an 80/20 basis, with a maximum provincial contribution of $400,000. Towns with a population between 5,000 and 15,000 could obtain shared funding on a 75/25 basis, with a maximum provincial contribution of $375,000 (Ministry of Community, Sport, and Cultural Development, n.d.).

References

Argent, N. 2011. "Trouble in paradise? Governing Australia's multifunctional rural landscapes", *Australian Geographer* 42(2): 183–205.

Bauch, P. 2001. "School-community partnerships in rural schools: Leadership, renewal, and a sense of place", *Journal of Education* 76(2): 204–221.

Beachy, T., Hanna, W., Ho, J., Nanton, D., Penney, W., and Renaerts, E. 2010. *Shared services: An opportunity for increased productivity.* A report of a project on shared services undertaken by the United Community Services Co-op on behalf of the Non-Profit Sector Labour Market Partnership. Vancouver, BC: United Community Services Co-op.

Blue-Howells, J., McGuire, J., and Nakashima, J. 2008. "Co-location of health care services for homeless veterans: A case study of innovation in program implementation", *Social Work in Health Care* 47(3): 219–231.

Browne, J. 2011. *Shared services survey 2011: The report.* London, UK: Browne Jacobson, Less Ordinary.

Bryant, C. 2010. *Co-constructing rural communities in the 21st century: Challenges for central governments and the research community in working effectively with local and regional actors.* Oxfordshire: CABI Press.

Charest, P. and Chanter, E. 2011. *Giscome (Willow River), Hixon, and Nukko Lake neighbourhood learning centres feasibility study.* Prepared by Generations Land Use Planning and Consulting. Prince George, BC: Regional District of Fraser-Fort George and School District No. 57. Available online at: http://sd57dpac.ca/wordpress/wp-content/uploads/2011/06/NLC-Feasibility-Study.pdf. Accessed on: January 5th, 2016.

Community Foundations of Canada. 2009. *Community foundation strategic alliances: Partnering for impact and sustainability.* A Discussion Paper for Community Foundations of Canada. Ottawa, ON: Community Foundations of Canada.

Conway, M.L., Dollery, B., and Grant, B. 2011. "Shared service models in Australian local government: The fragmentation of the new England strategic alliance 5 years on", *Australian Geographer* 42(2): 207–223.

Corbett, T. and Noyes, J.L. 2008. *Human services systems integration: A conceptual framework.* Discussion Paper No. 133-08. Madison, WI: Institute for Research on Poverty.

Dollery, B. and Akimov, A. 2007. "Are shared services a panacea for Australian local government? A critical note on Australian and international empirical evidence", *International Review of Public Administration* 12(2): 89–102.

Evans, P. and Grantham, B. n.d. *Friendship, courtship, partnership: Why Canadian nonprofits need to think about working together differently.* Milton, ON: Charity Village.

Farquhar, C.R., Fultz, J.M., and Graham, A. 2006. *Implementing shared services in the public sector: The pillars of success.* Report March 2006. Ottawa: The Conference Board of Canada.

Grant, G., McKnight, S., Uruthirapathy, A., and Brown, A. 2007. "Designing governance for shared services organizations in the public service", *Government Information Quarterly* 24(3): 522–538.

Graves, R. and Marston, H. 2011. *Seeking shared success: Business model innovation through mergers, affiliations, and alliances. Stories and insights from across the community foundation field.* New York: Community Foundations Leadership Team, Council on Community Foundations.

Grimsey, D. and Lewis, M. 2002. "Evaluating the risks of public private partnerships for infrastructure projects", *International Journal of Project Management* 20: 107–118.

Halseth, G. and Ryser, L. 2006. "Trends in service delivery: Examples from rural and small town Canada, 1998 to 2005", *Journal of Rural and Community Development* 1(2): 69–90.

Halseth, G. and Ryser, L. 2018. *Towards a political economy of resource dependent regions.* London: Routledge.

Halseth, G. and Sullivan, L. 2002. *Building community in an instant town: A social geography of Mackenzie and Tumbler Ridge, British Columbia.* Prince George: University of Northern British Columbia Press.

Hanlon, N. and Halseth, G. 2005. "The greying of resource communities in northern BC: Implications for health care delivery in already under-serviced communities", *Canadian Geographer* 49(1): 1–24.

Janssen, M. and Joha, A. 2006. "Motives for establishing shared service centers in public administrations", *International Journal of Information Management* 26: 102–115.

KPMG. 2014. *Perth area municipalities: Joint service delivery review, Appendix C guide to shared service arrangements for Ontario municipalities.* KPMG. Retained by the County of Perth.

Lennie, J. 2010. *Learnings, case studies and guidelines for establishing shared and collaborative service delivery in the non-government sector: Evaluation of the multi-tenant service centre (mtsc) pilots project.* Brisbane: Department of Communities, Queensland Government. Updated version of 2008 report.

Malone, L., and Anderson, J. 2014. "The right staffing mix for inpatient care in rural multi-purpose service health facilities", *Rural and Remote Health* 14(Online): 1–6.

Manson, D., Markey, S., Ryser, L., and Halseth, G. 2016. "Recession response: Cyclical problems and local solutions in northern British Columbia", *Tijdschrift voor Economische en Sociale Geografie* 107(1): 100–114. DOI: 10.1111/tesg.12153.

Markey, S., Halseth, G., and Manson, D. 2012. *Investing in place: Economic renewal in northern British Columbia.* Vancouver: UBC Press.

Minas, R. 2014. "One-stop shops: Increasing employability and overcoming welfare state fragmentation", *International Journal of Social Welfare* 23: S40–S53.

Ministry of Community, Sport, and Cultural Development. n.d. *Towns for tomorrow: Program guide.* Victoria: Province of British Columbia. Available online at www.townsfortomorrow.gov.bc.ca/docs/20111_towns_for_tomorrow_program_guide.pdf. Accessed on: January 5th, 2016.

Paagman, A., Tate, M., Furtmueller, E., and de Bloom, J. 2015. "An integrative literature review and empirical validation of motives for introducing shared services in government organizations", *International Journal of Information Management* 35: 110–123.

Rumball-Smith, J., Wodchis, W., Koné, A., Kenealy, T., Barnsley, J., and Ashton, T. 2014. "Under the same roof: Co-location of practitioners within primary care is associated with specialized chronic care management", *BMC Family Practice* 15: 149.

Rural/Urban Cost-Sharing Task Force. n.d. *Cost-sharing for success: A pro-active approach.* Edmonton, AB: Alberta Association of Municipal Districts and Counties and Alberta Urban Municipalities Association.

Ryser, L. and Halseth, G. 2012. "Resolving mobility constraints impeding rural seniors' access to regionalized services", *Journal of Aging and Social Policy* 24: 328–344.

Ryser, L. and Halseth, G. 2014. "On the edge in rural Canada: The changing capacity and role of the voluntary sector", *Canadian Journal of Nonprofit and Social Economy Research* 5(1): 41–56.

Ryser, L. and Halseth, G. 2017. "Opportunities and challenges to address poverty in rural regions: A case study from northern BC", *Journal of Poverty* 12(2): 120–141. DOI: 10.1080/10875549.2016.1141386.

Ryser, L., Markey, S., Manson, D., Schwamborn, J., and Halseth, G. 2014. "From boom and bust to regional waves: Development patterns in the Peace River Region, British Columbia", *Journal of Rural and Community Development* 9(1): 87–111.

Scottish Government. 2007. *Shared services – Guidance framework, December 2007.* Edinburgh: The Scottish Government.

Skinner, M., Joseph, A., Hanlon, N., Halseth, G., and Ryser, L. 2014. "Growing old in aging resource communities: Linking voluntarism, aging in place and community development", *The Canadian Geographer* 58(4): 418–428.

Sloper, P. 2004. "Facilitators and barriers for coordinated multi-agency services", *Child: Care, Health & Development* 30(6): 571–580.

Stein, F., Lancaster, M., Yaggy, S., and Schaaf Dickens, R. 2011. "Co-location of behavioral health and primary care services: Community care of North Carolina and the Center of Excellence for Integrated Care", *NC Medical Journal* 72(1): 50–53.

Sullivan, L., Ryser, L., and Halseth, G. 2014. "Recognizing change, recognizing rural: The new rural economy and towards a new model of rural service", *Journal of Rural and Community Development* 9(4): 219–245.

Van Belle, J. and Trusler, J. 2005. "An interpretivist case study of a South African rural multi-purpose community centre", *The Journal of Community Informatics* 1(2): 140–157.

Walsh, P. 2008. "Shared services: Lessons from the public and private sectors for the nonprofit sector", *The Australian Journal of Public Administration* 67(2): 200–212.

Whitfield, D. 2007. *Shared services strategic framework.* European Services Strategy Unit. New Castle, UK: Sustainable Cities Research Institute, Northumbria University.

Zeemering, E. and Delabbio, D. 2013. *A county manager's guide to shared services in local government.* Collaborating across boundaries series. Washington, DC: IBM Center for the Business of Government.

Part V
Moving forward

12 Emerging issues for new rural service and infrastructure models

Greg Halseth, Sean Markey, and Laura Ryser

Introduction

This edited volume has focused on the role of services and infrastructure as part of supporting sustainable economies and resilient communities in the rural and small town resource-dependent regions of developed economies. Drawing upon ten authorship teams, the individual chapters share stories of rural and small town services and infrastructure issues, and of innovations in the provision of those services and infrastructures from four OECD states. The collection itself is divided into three parts. The first focuses upon ways in which new services and infrastructure arrangements are being shaped by changing government policy. The second focuses specifically upon examples of innovative and new service arrangements. The third is on new and/or innovative infrastructure arrangements in rural and small town areas. Together, the contributions highlight both specific and general dynamics of rural service provision. We see how the contextual specificity of certain communities and regions creates unique conditions that foster innovation. More importantly for our purposes here, and when viewed as a collective, we see common sets of challenges and potential opportunities for rural and small town places and regions if appropriate policy attention and investments are made to support new models of service and infrastructure provision.

The central argument of the collection is that a more entrepreneurial approach to local service delivery and infrastructure provision by local organizations and local governments is needed to provide critical economic and community development supports, and also to assist in unleashing a more creative and innovative set of solutions that will both work locally and meet the needs of 21st-century rural and small town places and regions. Such an entrepreneurial approach, however, should not be seen as a surrender to the neoliberal trajectory of 'abandonment' approaches by so many senior levels of government, but rather as place-based and bottom-up initiatives that are nested within, and must be supported by, top-down supportive public policy. The community development foundations for resilient rural and small town places and regions must be co-constructed and co-delivered in partnership with both local and senior government actors in terms of both policy and committed fiscal resources.

In this concluding chapter, we draw from the previous chapters to expand upon our central thesis in the following four sections. The first section summarizes many of the challenges to providing rural service delivery. The second covers structural solutions to rural service delivery and infrastructure provision. These are the bigger picture: support mechanisms and approaches that will lead to more sustainable service delivery in rural locations. The third section covers many of the pragmatic innovations and stories seen throughout this volume in terms of rural service delivery. The final section covers some of the policy implications and includes policy recommendations.

The challenge of rural services

The rural context presents a range of challenges for the provision of rural and small town services and infrastructure. Large distances, small populations, and low population densities all increase delivery costs on a per capita basis. The now-40-year-long engagement with neoliberal public policy approaches means that the application of market-based delivery mechanisms and evaluation criteria exacerbate these base challenges.

Against these conditons of the rural context are the dynamics of contemporary rural change related to both the need for and provision of services. As described in the introduction and in a number of chapters, central governments have increasingly lost their points of contact with rural and small town places such that public policy contains more of an urban bias than in the past (see Blackburn; Nel and Connelly; Sherry and Shortall). In addition, the demography of rural and small town places in most OECD countries is following a similar transition towards resource frontier aging. When examined against the neoliberal policy trends in services provision, a curious and troubling disjuncture occurs. In many of these locations, the young populations of the 1960s and 1970s required relatively few medical care services yet their communities were equipped with full hospitals. Demographic aging now places the average population closer to 60 years of age in many settings, yet as the potential need for health care increases, the loss of local services means people face long drives during all seasons to adjacent regional centres for medical care. These changes mirror the challenges identified in the transition from space-based policy approaches to place-based policy approaches. The uniform application of regional development policies and servicing plans in the immediate post-World War II era has been undercut and new solutions require greater attention to the unique circumstances of places and the capacity of those places.

In addition to these general challenges of rural context and the dynamics of contemporary rural change, the specific contributions in this edited volume highlight five areas of challenge.

The first challenge highlighted across the contributed chapters builds from the small populations and limited human capacity of rural places. Authors highlight challenges associated with retaining and recruiting service professionals, but also with supporting, retaining, and recruiting residents as volunteers to participate in the direct provision of many services no longer operated or provided by the state (see

Gibson and Barrett; Hanlon *et al.*; Winterton *et al.*). These challenges of small population numbers are being exacerbated by demographic change, especially population aging. As service loads are increasing, existing services are being reduced, and many of the remaining services are dependent upon precarious volunteer support. All of these are critical challenges because of the vital role that services and infrastructure delivery play in building a robust platform for community development.

A second challenge area has to do with social issues around cooperation, competition, and partnership development. Individual chapters highlight the challenges of managing complex service delivery partnerships, often amongst volunteer-organized groups, at both the local and regional levels (see Dollery; Hanlon *et al.*; Ryser *et al.*). Other chapters highlight that inter-community competition can make regional cooperation a challenge, even when all parties recognize that a united regional voice can have a greater impact in resolving collective infrastructure and service delivery needs.

We also note that the pursuit of regional solutions is often challenged by a backdrop of regionalization. Regionalization refers to situations where senior government policy decisions seek to rationalize costs through centralization of services or service infrastructure. This means that communities may suffer losses of services in favour of centralized service delivery where one community benefits and others lose. The principles of regionalism in terms of collaborative action, coordinated resources, and community specialization are taking place within a political setting where regionalization is more often the outcome, which exerts negative pressure on the potential for positive inter-regional collaboration.

A third area of challenge identified across the contributions has to do with the broader policy and political context. The transition from a Keynesian to a neoliberal public policy approach has included significant offloading of programs and service delivery to the local level (see Gibson and Barrett; Ryser *et al.*; Sherry and Shortall; Winterton *et al.*). Such offloading has not been accompanied by funding supports equivalent to the delivery of those services so that services are either not delivered or are being delivered by local voluntary groups in less robust or institutionalized ways. Against this backdrop, and related to the previous point above, service delivery may be further complicated when services are not offloaded, but instead are regionalized or centralized, making them potentially even more inaccessible to rural residents.

A further element associated with the political challenges of service delivery concerns the role of local governments. Local governments are often challenged under a neoliberal policy framework to do more, but the fiscal and regulatory regimes within which they work have not been significantly retooled to allow these local governments to meet that challenge. In response, we see cases where communities are scrambling with short-term and 'quick fix' solutions, which may present a patchwork solution to maintain service delivery in the short term but may create longer-term challenges to more robust, transformative change.

A further aspect of political and policy challenge concerns inactivity in both rural policy development and the implementation of supports following investigations into rural service needs. In this case, hollow commitments to equity across geographies, or to the application of action following needs identification, are challenges that enhance feelings of powerlessness and fatigue on the part of rural service and infrastructure advocates (see Hanlon *et al.* and Sherry and Shortall).

The fourth area of challenge identified across contributing chapters focuses upon the built environment and infrastructure services of rural and small town places. Principal among identified challenges is the persistence of a deep digital divide between rural and urban locales. The market-based application of new information technologies confronts the base challenges of rural service provision outlined above with the outcome being that rural and small town places continue to fall behind in their access to information technologies. Given that so much of the contemporary economy is not just about moving goods or people but also about moving information and ideas, this persistent and growing gap is a significant challenge. There are exceptions, and there are significant policy interventions, but these do not make up for the pace of change in new information technologies being driven in the private sector through urban nodes (see Dollery; Kelly and Hynes).

Built infrastructure challenges also include the growing rural infrastructure deficit. The failure to invest through decades of neoliberal policy adjustment and state fiscal 'belt-tightening' means that rural infrastructure has been degraded and is often in the last stages of the infrastructure life cycle (see Minnes *et al.*). Complicating this gap is the fact that efforts to create investment tools and mechanisms to renew rural infrastructure have struggled to find supporters when confronted with similar infrastructure deficits in urban locations, which get prioritized. Similar to the digital divide outlined above, a crumbling rural infrastructure does not support the needed community development or the community economic development foundations of the 21st-century rural economy.

A final area of challenge identified across the chapters concerns economic and financial barriers. Fiscal issues have been woven throughout the previous four challenge areas. First among these is that rural and small town locations are not conducive to market-led service provision solutions, even where subsidies and incentives might be made available through policy mechanisms. The application of private sector models in provision or evaluation does not fit the contemporary context or circumstances of rural and small town places. Where economic and financial resources are being made available to economic and industrial interests, the failure of supports in the other challenge areas identified above means that those investments will not realize their full potential. New rural businesses or industries may be hampered by underdeveloped and crumbling physical infrastructure, a persistent digital divide, the limitations of social and partnership relations within communities and across regions, and the limited human resources and human capacities of small places.

Rural investments in service delivery need to be well coordinated across a wide spectrum of policy areas.

Structural solutions to rural service delivery

In this section, we outline some of the structural conditions and approaches that are necessary to ensure the durability, sustainability, and even potential expansion of the innovative examples seen throughout the chapters. The first concerns adopting a place-based approach to service delivery within an embedded commitment to community development. This means that we must not view service and infrastructure provision in isolation. A place-based community development perspective facilitates a more integrated approach to service delivery. In this context, community development refers to the inter-connectedness of various forms of community capital (i.e., human, social, cultural, natural, political, built, and financial) that can be mobilized to strengthen the capacity of place-based resilience to economic transition and development (Halseth and Ryser, 2018). A community development approach enables service delivery organizations, agencies, and policy makers to understand service needs and the impacts of decisions (and decisions in different government service areas) in more contextualized terms.

Second, regional development, as shown in the chapters, is clearly part of the rural service delivery solution. We have already covered many of the challenges identified when 'regionalism' is pursued as 'regionalization'. Here regional development specifically addresses many of the structural conditions that hinder rural areas. Regional development offers many potential benefits, including collaboration to address issues of capacity; coordination to avoid duplication; and the pooling of resources in order to enhance the physical viability of different practices, services, and pieces of infrastructure.

Third, innovative service delivery demands new approaches for how to effectively manage human resources. This involves issues of retention, training, and effectively utilizing volunteer capacity in rural areas (see Ryser *et al.* and Winterton *et al.*). The goal is to leverage people's commitment to place and volunteerism as a value-add to service delivery, rather than as a last-resort option to retain critical services.

Finally, in a theme we will revisit further in the policy section below, governments and relations between different levels of government must operate under conditions of co-construction or, specific to rural service provision, co-delivery to maximize efficiency, combine resources, and adopt appropriate place-based solutions. Co-construction seeks to facilitate and operationalize innovative and durable solutions that combine the resources, expertise, and differing levels of jurisdictional authority between senior and local governments, and service organizations/ agencies to create, support, and invest in rural communities and regions. Such co-construction draws upon continued strategic engagement of senior governments as crucial partners who are equipped with the jurisdictional responsibility of many public policy levers that can support local and regional transformative changes across broad rural landscapes (Reimer and Markey, 2008).

Pragmatic innovations in rural services

We now turn our attention from the structural to the specific, highlighting examples of rural service innovation found within the chapters that we hope will inspire other jurisdictions. One of the common innovations in rural service delivery focused upon the creation of streamlined one-stop shops. Such consolidated service provision hubs provided single access to multiple service points and mechanisms. Depending upon the user group and the context, such service provision hubs may have been physical or live delivered, while others might have been coordinated in a virtual manner. In all cases, these hubs reduced the overhead costs for individual service providers and the transaction costs for service users (see Blackburn; Ryser *et al.*).

A second body of pragmatic rural service delivery innovations focused upon those which enhanced the entrepreneurial supports and tools available to communities and regions in order to retain, and even extend, rural service and infrastructure delivery. Examples across the chapters included things such as health trusts, community foundations, and upstream and downstream service support renovation (see Hanlon *et al.*; Nel and Connelly). In all cases, these innovations in entrepreneurial tools were specifically aimed at supporting resilient and sustainable community development in small places. While these forms of support have historically played important community development functions in communities, supporting various causes or issues, we see here a shift to use these mechanisms to deliver core services in new ways.

A third grouping of pragmatic innovations focused upon place-based approaches to building community capacity. Recognizing the challenges of small populations but also the inherent strength of the strong ties between individuals and groups in such settings, these innovations sought to take advantage of programs such as 'train-the-trainer', scaling up specialization, supporting families and individuals to take their part in the service delivery chain, and working with community leaders to support broader community-based programming and infrastructure management (see Dollery; Minnes *et al.*; Winterton *et al.*).

A fourth area of innovation focused upon the testing of new and creative models of infrastructure arrangements. These included experimenting with cost saving or management-focused software, as well as physical decisions about co-location or relocation of units in order to be more effective and to support dialogue and innovation. The testing of innovative models provided opportunities to explore the potential for greater flexibility in service and infrastructure provision, to explore the trading off of different levels of service delivery at different times and to different client groups, and to overcome structural limitations in service and infrastructure delivery. In each of these cases, the testing of new and innovative models built upon a much more integrated vision of approaching community and economic development support (see Dollery; Kelly and Hynes; Nel and Connelly; Ryser *et al.*).

Across all of the innovations profiled in the chapters in this book, there are a range of key features common to most. First, policy makers and agencies need to make sure that the service or infrastructure model fits with the rural and small town context. This means not simply transposing urban models to a context where they will likely not fit or work. A second key feature focuses upon cooperation, partnering, and relationship building in order to reduce overhead costs for providers and transaction costs for clients. Third, the examples here all involve the creative re-imagining of the service itself, of service production, and of service delivery. This means that policy makers and agencies are not simply seeking to restore conditions of the past, but are actively engaged in under-standing local and regional conditions and thinking creatively about how to effectively use assets to deliver the services that are needed to rural communities. Finally, most of the examples also focus upon processes for scaling-up capacity building and building capacity locally. We must move beyond an endless cycle of pilot programs and move more quickly to deliver proven models.

Perhaps the most interesting common theme in innovation seen in the cases was the recognition that 'place' itself was a stepping-off point for innovation. In this case, efforts sought to understand how the unique assets of place could be mobilized to address a specific need or a gap unique to a particular setting, and which drew upon existing strengths and potential. A second body of innovation focuses upon the 'wise' application of technology to enhance service delivery. Too often 'smart service delivery options' are simply the replacement of local service delivery with services or information now made available over a website. Instead, the innovation concerning the use of technology presented a more sensible or wise application to meet the needs of the community and the client group, as well as to employ technologies to, again, reduce overhead or transaction costs in order to continue to allow the services to exist locally. Finally, most of the examples drew upon innovative practices that focus upon co-management, co-governance, co-location, and/or the co-delivery of services. In other words, it was all about breaking down the walls (literally and figuratively) that had been built in the 1950s, 1960s, and 1970s era of service provision in order to re-organize into an integrative framework that makes sense for the 21st century.

Each of the cases is looking for innovative solutions to service delivery in rural and small town places that are confronting the needs and demands of a 21st-century global economy in which economic development and community development go hand in hand. They are not divisible. Service provision in one sector will not be of assistance unless it is complemented by service provision in the other sector. Taken together, all of the pragmatic innovations shown in the chapters emphasize a focus upon building resilience and sustainability at a local level, in local services and infrastructure, and in the local community and economy itself.

Policy

Emerging out of 40 years of neoliberalism, where the primary policy objective was to not have policy and to let the market decide, we are faced with a variety of inherent contradictions within the neoliberal approach as applied to rural and small town places. This means that while senior governments offloaded responsibilities (in the hopes of unleashing local or regional entrepreneurial capacity and responses), it was done without thought, funding, or capacity to enable a variety of supportive conditions. Offloading or state withdrawal has occurred without transferring jurisdiction (meaning that senior governments maintained responsibility), without any transfer of resources, even resources to build capacity in transition, and during a time that coincided with the end of the life cycle of much service infrastructure. Perhaps most problematic is that state withdrawal over this period coincided with the conditions and trajectory of the global economy where local services and amenities matter considerably more to the economic development prospects of rural and small town communities and regions than they did in the past.

In seeking to create a more appropriate policy response, we do not seek to turn back the clock to dependence on senior governments. Clearly, the solutions contained within this volume provide a window into the innovative capacity of communities to deliver services that are well matched to local conditions (see summary Table 12.1). There is also the reality that communities are now more engaged and responsible for service delivery – in other words, power they are unlikely to relinquish. What we see in the chapters are examples that show a new way forward, where there are responsibilities for communities, regions, and senior governments. This co-constructed future will rely on a new approach to policy development and implementation.

Conclusion

Rural and small town places in developed economies have faced decades of social, economic, political, cultural, and demographic restructuring. Against this rising set of challenges, these same rural and small town regions have experienced limitations in their local capacity and the withdrawal of senior government and industry supports to address and manage transformative change. Within this context of change and transformation, the provision of rural services and infrastructure is fundamental to both rural community development and rural economic development.

The contributions within this book have all described aspects of rural and small town service and infrastructure provision. The authors have highlighted challenges and innovative responses. Taken together, the chapters highlight a great deal of commonality in challenges and a great deal of creativity in innovative responses. In this concluding chapter, we have brought together these common understandings and examples, and have summarized key features and critical policy implications in order to inform wider debates and actions around rural and small town service provision and infrastructure delivery in a wide range of developed economy contexts. The future of these rural and small town places and regions can be very bright through the opportunities now being created within the new rural economy, but to realize those new and exciting futures, we must attend to the fundamentals of service and infrastructure provision.

Table 12.1 Policy innovations

Policy Innovation	Chapters
Rural proofing: Rural proofing and mainstreaming rural policies provide a mechanism to enhance senior government knowledge of rural conditions.	Sherry and Shortall
Scale-up pilot programs: Rural communities advocating to more quickly scale up learning and innovations from pilot programs in order to avoid an endless cycle of pilot projects and short-term funding.	Hanlon *et al.*; Minnes *et al.*
Local governments as conveners: Local governments serving as conveners of place-based policy to connect local service organizations and bring senior governments to the table to support innovative solutions.	Dollery; Kelly and Hynes; Nel and Connelly; Ryser *et al.*
Re-scale financial resources and jurisdiction: Policies designed to re-scale financial resources and issues of jurisdiction which may otherwise be preventing the longevity of innovative rural solutions.	Gibson and Barrett; Minnes *et al.*
Pragmatic accounting and monitoring: Communities and organizations advocating for more sensible accounting and monitoring frameworks to avoid overburdening local capacity, yet retain key learnings for continuous improvement.	Dollery; Ryser *et al.*; Winterton *et al.*
Life cycle approach to human capacity: Small communities pursuing policies designed to support a life cycle approach to capacity and skills development.	Nel and Connelly; Winterton *et al.*
Synchronize infrastructure investments: Rural communities seeking policy and investment approaches that synchronize infrastructure investments (i.e., expansion of broadband with highway and railway infrastructure investments).	Kelly and Hynes
Adopt value-added regional solutions: Communities and senior governments pursuing strategic policies to convene and incentivize regional solutions, including attention to how senior governments can play a role in positive regional development responses.	Blackburn
Integrated senior government policy: Policies are needed to support coordination across senior government agencies so that innovative approaches are not overwhelmed or cancelled out by other realms of government. This includes attention to building top-down capacity in order to understand, facilitate, and evaluate rural responses.	Blackburn; Sherry and Shortall; Ryser *et al.*

References

Halseth, G. and Ryser, L. 2018. *Towards a political economy of resource-dependent regions*. London: Routledge.

Reimer, B. and Markey, S. 2008. *Place-based policy: A rural perspective*. Report to Human Resources and Development Canada. Accessed April 23, 2018. Available online at www.crcresearch.org/files-crcresearch_v2/files/ReimerMarkeyRuralPlaceBasedPolicySummary Paper20081107.pdf.

Index

Printed and bound by CPI Group (UK) Ltd, Croydon, CR0 4YY

24/10/2024

01778306-0012